GOING SOUR
Science and Politics
of Acid Rain

D1262540

GOING SOUR

Science and Politics of Acid Rain

ROY GOULD

BIRKHÄUSER BOSTON • BASEL • STUTTGART

Library of Congress Cataloging in Publication Data

Gould, Roy, 1947–
  Going sour.

  Bibliography: p.
  Includes index.
  1. Acid rain — Environmental aspects — United States.
  2. Acid rain — United States — Political aspects.
  I. Title
  TD196.A25G68   1985   363.7'386'0973   84-21700
  ISBN 0-8176-3251-4

CIP-Kurztitelaufnahme der Deutschen Bibliothek

Gould, Roy:
Going sour: science and politics of acid
rain / by Roy Gould. – Basel ; Boston ;
Stuttgart : Birkhäuser, 1985.
    ISBN 3-7643-3251-4

37–38 : 44   31–38 : 44

©Birkhäuser Boston, 1985
ABCDEFGHIJ
ISBN 0-8176-3251-4
ISBN 3-7643-3251-4
Printed in the United States of America

To RSG and the memory of my father.

ACKNOWLEDGMENTS

This book had its origins in a scientific review and policy analysis of acid rain that I undertook while a Fellow at the Interdisciplinary Programs in Health at the Harvard School of Public Health. I am grateful to many colleagues for stimulating discussions of the science of acid rain, and would particularly like to thank Harvey Brooks, John Cairns, Don Hornig, Jennifer Logan, Peter McKinney, Rashid Shaikh, John Spengler, George Thurston, Ton Schoot-Uiterkamp and Steve Wofsy. Alas, any errors are entirely mine. The views expressed in this book are solely those of the author.

Portions of Chapters 2 to 6 have appeared in *The Annual Review of Energy.* 9:529-559. (1984).

TABLE OF CONTENTS

PART 1 INTRODUCTION

Chapter 1 ACID RAIN: AN OVERVIEW OF THE ISSUES 3

PART 2 THE SCIENCE OF ACID RAIN

Chapter 2 THE COMPOSITION OF RAIN 43
Chapter 3 THE CAUSES OF ACID RAIN 46
Chapter 4 THE TRANSPORT OF ACID POLLUTION 55
Chapter 5 THE EFFECTS OF ACID RAIN 63
Chapter 6 STRATEGIES FOR REDUCING ACID RAIN 81

PART 3 POLITICS AND ACID RAIN

Chapter 7 ACID RAIN AND THE POLITICS OF SCIENCE 89
Chapter 8 ACID RAIN AND THE LAW 114
Conclusion 123

APPENDIX

Plates 129
Tables 132
Figures 136
Glossary 143
References 145
About the Author 155

PART 1

INTRODUCTION

1

ACID RAIN: AN OVERVIEW OF THE ISSUES

"If there is life and death in the air, we must believe the same of rain . . ."
R.A. Smith, 1872

"Right as rain" and "pure as the driven snow" are expressions from a bygone era. Now the storms that sweep across eastern North America carry an acid rain — a rain gone sour. Tainted by pollution from the burning of fossil fuels, the rain is no longer "a kind physician":

• In New York State's Adirondack Mountains — and elsewhere in the Northeast and Canada where the underlying soil and rocks have been unable to neutralize acid rain — hundreds of once-pristine lakes and streams have gradually grown acidic, and the aquatic life they sheltered has dwindled and vanished.

• On the high ridges of the eastern U.S., from the Green Mountains of Vermont southward to the Great Smoky Mountains of Tennessee, the morning mist may be as much as 100 to 1,000 times as acidic as unpolluted mist. Along parts of this mountainous spine, majestic stands of red spruce have virtually ceased growth.

• In communities in the Northeast, water supplies have grown acidic and "aggressive," dissolving lead and other toxic metals into drinking water.

• From the tip of Louisiana to the coast of Maine, the summer months bring an acid haze that can decrease visibility to less than a mile. As far away as Bermuda, nearly 600 miles offshore, the level of sulfuric acid in the air shoots up six-fold when air masses from the eastern U.S. arrive with the wind.

• No longer is the problem confined to eastern North America. In the Colorado Rockies, high mountain lakes are undergoing the first stages of acidification. In Southern California, the fog is sometimes as tart as the acid in a car battery. Acid haze from three continents travels as far north as the Arctic Circle, where the once-pristine snow now bears the acid imprint of man-made pollution.

Acid rain has emerged as the most important and controversial environmental problem of the decade. It is the first global-scale problem to be caused by burning fossil fuels, and it marks a serious turning point in our brief history on the planet. Unlike the toxic wastes at Love Canal or the radioactive hazards at Three Mile Island, the substances that cause acid rain — sulfur and nitrogen oxides — are found throughout nature and are in fact part of nature's essential plan. But the burning of fossil fuels has released these substances into the air in quantities that dwarf nature's own output, and that for the first time have upset the balance of nature on a planetary scale. A related problem of this new type is just now emerging from the wings: the so-called "greenhouse effect," which threatens to alter the earth's climate. Like acid rain, the greenhouse problem is caused by the release of a simple substance — carbon dioxide — from the burning of fossil fuels, on a scale that has upset the balance of nature.

Pollution is an acquired taste. Garbage is not to everyone's liking, and it is not surprising that at the very moment in history when garbage has become a central theme, we relegate it to the "back burners" of our attention. We are inundated by reports of environmental woe and we are inured to the consequences. But the story of acid rain is not just a tale of two more pollutants. It is a drama that weaves history, scientific sleuthing, political intrigue, and that most interesting of all mysteries, human nature.

Acid rain is a litmus test for the nation, a test of how well we choose to confront and solve our environmental problems. To date, the public debate over acid rain has gone as sour as the rain itself. No other controversy in recent times has so divided the nation along regional lines, nor engendered such bitter dispute among powerful competing interests. Much of this book will examine whether the public interest has been served by the institutions we rely on to resolve scientific disputes and to formulate environmental policy: the Environmental Protection Agency, the Congress, the scientific community, the press, and the courts. We will argue that the public interest has *not* been well served, and that beyond the immediate issue of acid rain, something ugly and potentially disastrous has been happening: the science of acid rain has been increasingly misrepresented and distorted for political ends. The scientific community has felt the chilling effect of politics to a degree unprecedented in recent times. Acid rain will have had a silver lining if it forces us to confront and to reverse this ominous trend toward the politicization of science.

This chapter presents an overview of the acid rain problem. What were the roots of the problem and how did they grow? How has acid rain affected the environment? What are the social, economic, and political consequences of cleaning up acid rain? How have industry and government chosen to respond to the challenge? Chapters 2 to 6 document in more detail the science of acid rain, with emphasis on the causes and effects. Readers more interested in the public debate on acid rain may skip directly to Chapters 7 and 8 and refer to the science chapters as needed. Chapter 7 examines the ways in which the public's understanding of acid rain was shaped — and often misshaped — by the major actors in the acid rain debate. Chapter 8 briefly examines the role of the Environmental Protection Agency and the courts in administering the laws governing air pollution.

The problem in brief

Acid rain is caused by sulfur dioxide (SO_2) and nitrogen oxides (NO_x) released to the air during the burning of coal, oil and other fossil fuels.* Within hours to days, these pollutants are oxidized in the air to form acid sulfate and acid nitrate — commonly known as sulfuric acid and nitric acid. The acids are brought to earth in rain, snow and other forms of precipitation such as dew, mist and frost. Even when the sun shines, microscopic particles of acid sulfate and acid nitrate continually trickle to earth as "dry deposition," a kind of acid fallout which adds as much acidity to the environment as acid rain itself. The various pathways by which acid returns to earth are collectively termed "acid deposition," but this book will use the popular term "acid rain" except where confusion would arise.

The continual shower of acid extends over most of eastern North America, from the Gulf of Mexico to northern Canada and out over the Atlantic Ocean. About two thirds of the acidity is acid sulfate, while one-third is acid nitrate. The most seriously affected regions are the northeastern U.S. and Ontario, Canada, which not only receive the most acidic rain but are also the most sensitive to acid rain, because their underlying rocks and soils cannot neutralize the incoming acidity. Other sensitive regions include parts of the Appalachian Mountains, the Great Smoky Mountains, the Boundary Waters Canoe Area in northern Minnesota, and parts of the Rocky Mountains. However, there are pockets of susceptibility throughout the eastern U.S. Furthermore, the direct effects of airborne acid may be felt in a very broad region.

We all contribute to acid rain every time we light a match, drive the family car, or burn any fuel. However, most of the pollution that causes acid rain comes from just a few large sources. Nearly three quarters of the SO_2 emitted east of the Mississippi River comes from power plants, and nearly half of that from just the forty largest coal-fired power plants. A single one of the largest coal-fired plants, such as the Muskingum plant

*Nitrogen oxides are a mixture of nitric oxide, NO, and nitrogen dioxide, NO_2, collectively abbreviated NO_x, where x = 1 or 2.

in Ohio, annually emits about as much SO_2 as Mt. St. Helens' volcano — some 200,000 tons a year.

The largest plants use very tall smokestacks, some of which rival the Empire State Building in height. SO_2 from these "superstacks" can be carried hundreds or even thousands of miles downwind, remaining aloft long enough for a considerable fraction of the SO_2 to be converted to acid before returning to earth. Thus the largest plants contribute preferentially both to the formation of acid rain and to the transport of air pollution across state and national boundaries. The 50 largest coal-fired power plants are mostly clustered in the Midwest and the Ohio and Tennessee Valleys, and are currently the focus of Congressional efforts to reduce acid rain. SO_2 is emitted from many of these plants without any pollution controls.

The sources of NO_x emissions are more widely distributed. About 44% comes from transportation (motor vehicles and, to a lesser extent, trains and planes). The remaining 56% of NO_x emissions comes chiefly from the smokestacks of power plants and other industrial sources. A small percentage comes from homes and businesses. Although Congressional attention has focused on SO_2 emissions — the major cause of acid rain — NO_x contributes a significant and growing share of the acidity in rain.

Every state contributes to its own acid rain. However, large amounts of air pollution are carried between states. For example, New York's Adirondack Mountains, an area hard-hit by acid rain, is downwind of the major sulfur emitters in both the Midwest and Canada. According to a variety of studies, including computer calculations and actual field measurements, more than 50% of the acid sulfate in the Adirondacks comes from Midwestern SO_2 emissions, about 20% comes from the large metal smelters in Ontario, Canada, while less than 10% comes from the Northeast itself. Thus even if the Northeast completely eliminated all of its SO_2 emissions, there would be little impact on the amount of acid rain in the region.

But the inequity between regions goes even deeper. Northeastern power plants burn relatively clean oil, in contrast to the cheap, high-sulfur coal burned by midwestern plants. As a result, northeasterners pay up to 4 times more for their electricity. Midwesterners pay an artifically low price for electricity, because part of the real cost of producing electricity — dealing with the consequences of pollution — has been shifted to north-easterners and other parties downwind. Under even the most costly plans for reducing SO_2 emissions, midwesterners would still pay less for electricity than do northeasterners.

The technological solutions to acid rain are at hand, but the political obstacles have been formidable. The midwestern electric utilities and the coal industry have formed a powerful coalition to block legislation aimed at cleaning up acid rain. It is not hard to see why: Reducing SO_2 emissions from power plants would increase the cost of generating electric power. It would also reduce the demand for high-sulfur coal, leading to a decline in production and the loss of thousands of jobs in the industry. The battle lines have been clearly etched between the Midwest and the Northeast.

Another storm has been brewing to the north. The United States sends Canada more than 5 million tons of airborne acid sulfate annually — about 20% of the sulfur emitted in the eastern U.S., and two to three times more sulfur than the U.S. receives from Canada. Considering that Canada has more resources at risk to acid rain than any other nation in the world, it is not surprising that Canada's Minister of the Environment, John Roberts, described acid rain as "the single most important issue" between the two nations. Canada has taken the initiative on acid rain, challenging the U.S. to a mutual 50% reduction in SO_2 emissions.

A balance between fire and life

To appreciate the turning point to which acid rain has brought us, it helps to step back a few million years to get a panoramic view of the natural cycles of sulfur, nitrogen, and acidity. Throughout the earth's history, nature has created her own weak brew of acid rain. Much of this acid originates with fire. For example, forest fires, lightning bolts, and volcanoes spew forth SO_2 or NO_x. Once in the air, these compounds combine fully with oxygen to form acid sulfate and acid nitrate. There is nothing mysterious about the process: in nature, acids are often formed as a by-product of oxidation. In fact, the word "oxygen" literally means "acid-former." (This is painfully familiar to anyone who has left butter exposed to air for too long and has tasted the acids responsible for rancidity.)

Acid sulfate and acid nitrate are extremely stable chemically; if there were no way to get rid of them, they would gradually accumulate and turn the earth's lakes and rivers into giant pickling baths of acid. Fortunately, living things such as plants and bacteria can convert acid sulfate and nitrate back to the chemically "reduced" forms needed to make proteins and other molecules essential for life. This reverse process *consumes* acidity. The entire biosphere is dependent on organisms that can rescue sulfur and nitrogen from their most oxidized state and return them to a form essential for life.

As sulfur and nitrogen are cycled through the environment and shuttled between their oxidized and reduced forms, acidity is continually produced and consumed, and a balance is reached. It is a curious balance between "fire" and "life," a balance that has been stable over millions of years — until now.

Our species has been releasing SO_2 and NO_x into the air ever since we discovered fire and came in from the cold; yet for millenia, our emissions of these gases were barely a ripple on nature's own output. During the past century, however, the scale of the human enterprise grew immensely larger. The population grew exponentially, and the engines of progress demanded increasing amounts of energy. To stoke the fires of the Industrial Revolution, we brought coal to the surface of the earth by the billions of tons. Sulfur and nitrogen that had been locked in the coal for millions of years were released to the atmosphere within just a few generations — a bare instant from the geological perspective. Today, man-made emissions of SO_2 and NO_x equal the earth's entire natural output. In the eastern U.S., the tens of millions of tons of SO_2 and NO_x we pour into the air each year dwarfs nature's own emissions within the region by ten to fifty times. As a result, we are showering the environment — day and night, rain or shine — with unprecedented acidity. As we shall see, nature's normal mechanisms for coping with acidity have been overwhelmed. The balance between "fire" and "life" has been tipped in favor of fire.

"Acid rain is scarcely a decade old . . ."

The warning signs of acid rain went largely unheeded for more than a century. As early as 1838, the mayor of Boston, Samuel Eliot, noted in his inaugural address that "the supply of rainwater is more and more affected by the increased consumption of bituminous coal," an issue he called "the most important" facing Boston's City Council[1]. Fearful of interfering with the growth of textile mills and other industries, Boston chose not to burn coal more cleanly, but instead to find more pristine sources of drinking water. Today, a century and a half later, Boston draws its water from the most pristine part of the state, the huge Quabbin Reservoir, which now receives among the most acidic rain in the eastern U.S. Like other lakes and streams in the eastern U.S., tributaries to the Quabbin's west branch have recently been found incapable of supporting trout and other game fish[2]. Evidently we no longer have the luxury of merely seeking new resources when the old ones have been polluted.

By the mid-nineteenth century, acid rain was in full flourish in Britain. The stage had been set in the time of Queen Elizabeth I, when coal began to replace scarce wood as a fuel. The increasing pall of pollution caused by coal burning from the 1600s on can be seen in the changing colors of the background sky in landscape paintings from the period — the colors progressing from blue, to pink, to the muddy yellow-browns which inspired the term "pea-soup fog."[3] The pollution was so bad that it became difficult for a person to see his feet while walking, and lamps had to be used throughout the day. Sunlight was so diminished that the incidence of rickets increased dramatically; by the end of the nineteenth century, half the children in the poorest districts of Leeds were suffering from it.[4] Today, smoke from coal combustion is removed before it leaves the smokestack, but the invisible pollutants — SO_2 and NO_x — are still spewed from many coal-fired plants just as they were a century ago.

In the days when Charles Dickens was vividly describing the horrors of the Industrial Revolution, a now-forgotten figure was painting an equally chilling "chemical portrait" of the landscape. In his landmark book, *Air and Rain: The Beginnings of a Chemical Climatology*, the British chemist Robert Angus Smith coined the term "acid rain," and summarized twenty years' of measurements on the chemistry of rain in Britain and Germany[5]. The year was 1872. Smith, the father of environmental chemistry, was one of the leading scientists of his time — Fellow of the Royal Society, Inspector-General of the Alkali Works for the British Government, and one of the last students of the great German chemist, Justus von Liebig, to whom he dedicated *Air and Rain*. Smith was capable of the most meticulous calculations and the most impetuous experiments. In order to determine the effect of carbon dioxide on respiration, he once sealed himself and a candle into a lead-lined vault to see which would expire first. Just before collapsing, "I suddenly realized my own life was but a brief candle," he wrote, "and I was seized with a burning desire to *live.*"

With the aid of friends and colleagues throughout Britain, Smith set up the world's first network for collecting and analyzing samples of rain. He admonished his colleagues to "obtain specimens of rain-water exactly in the condition in which it falls . . . It is difficult in words to convey an idea of the importance in such experiments of obtaining freedom from all extraneous substances." To insure that his own rain samples not be contaminated, he collected some of them in a platinum bowl which had been heated red-hot to remove all traces of fingerprints and other impurities. An excerpt from his laboratory notebook: "Nov. 27, 1869 . . . Rain obtained from: Edward Street, Hampstead Road, close to Cumberland Market, during a gale . . . Wind: WSW . . . Sulphuric acid: 1.4700 grains per gallon (21 parts per million) . . ."

Smith discovered that the acidity of rain "is caused almost entirely by sulphuric acid . . . and in country places to a small extent by nitric acid and by acids from combustion of wood, peat, turf, etc." In industrial centers such as Liverpool and Glasgow, the rain was extraordinarily acidic — containing some twenty times as much acid sulfate as it does today — "because of the amount of sulphur in the coal used." (See Appendix.) In fact, rain would have been even more acidic had it not also contained very high levels of *alkaline* pollutants, such as ammonia from animal wastes. (As the French chemist J. I. Pierre put it in the 1860s, "This large amount of ammonia explains why . . . the fog is endowed with such a penetrating odor as to affect very sensibly the organs of respiration.") The acidity of rain was also partly neutralized by huge quantities of alkaline fly-ash — the grimy residue from coal combustion spewed into the air from smokestacks. Nevertheless, in the war between acid and alkaline pollutants, the acidity prevailed.

"The presence of free sulphuric acid in the air," Smith wrote, "sufficiently explains the fading of colours in prints and dyed goods, the rusting of metals, and the rotting of blinds." He observed that "the lower portions of projecting stones in buildings were more apt to crumble away than the upper; as the rain falls down and lodges there, and by degrees evaporates, the acid will be left and the action on the stone much increased." He also conjectured that the high mortality in Glasgow was due in part to the city's extremely poor air quality, as reflected in the chemistry of Glasgow's rain.

Acid rain and related forms of air pollution apparently caused serious damage to trees, crops and plants on the outskirts of cities. "Mosses may be seen to grow in the acid-rain of towns when trees, shrubs and grasses disappear." Smith grappled with many of the same problems that researchers face today: Was the damage due to acid rain or to the direct action of gaseous pollutants or natural factors? "Acid gases," such as SO_2 and hydrogen chloride, were known to be toxic to crops such as wheat; but Smith noted that acid rain damaged roots, while the gases did not. He also observed that grass and plants were protected by the wind-shadow of hedges, and he reasoned that the "shadow effect" would not be seen if the damage were caused by gases. He concluded that the damage was due to acid-laden droplets of mist driven by the wind. He recognized that trees and shrubs might simply be killed by violent storms, or by the salt from sea-spray — but in that case damage would be expected most severe along the coast. Yet the damage to vegetation was observed to increase inland, toward the cities, "in line of Liverpool and Prescot." Smith's work was remarkably prescient. At a time when microorganisms were still mysterious, he postulated that the discoloration seen on bark and leaves might be caused by fungi that attacked a plant as it was weakened by acid rain. He also conjectured that acid rain might interfere with the microorganisms reponsible for natural decay processes — a concern that has been voiced once again by contemporary scientists.

Britain obviously survived a century of acid rain, so why the concern today? The problem currently facing eastern North America and western Europe differs in several crucial respects from Britain's bout with acid rain in the 19th century. British soil is rich in limestone and other alkaline minerals that neutralize acidity as soon as it hits the ground (the chalk cliffs of Dover are a famous example). Britain is thus immune to many of the long-term effects of acid rain. Furthermore, the brunt of acid rain was largely confined to cities and their outskirts (Appendix). As we shall see, air pollution in the cities and industrial centers was gradually reduced by building ever taller smokestacks. Today, Britain sends much of its SO_2 far downwind to sensitive regions in Scandinavia and elsewhere in Europe.

Smith was both a scientist and policy-maker. As Inspector-General of the Alkali Works (a position equivalent to our chief of air pollution control), he grappled with the same problems of science, commerce, and public policy that beset us today. He concluded that in regions suffering from air pollution, "it seems quite fair to interdict the further extension of manufactories, unless, of course, these are carried on in an improved manner." .Smith's writings emanate the enthusiasm of an optimist who believes that society's ills can be cured by the patient application of science. Yet he acknowledged that the issues were "not simple, and the rules for judging the (environmental) condition of a place are not clear. I hope this will soon cease to be the case . . ."

He would have been disappointed. More than a century later, the current Inspector-General of the Alkali Works described himself as "not concerned with the sulfur in Scandinavia." The chief of environmental studies at Britain's Central Electricity Generating Board could say curtly, "When people flick their light switches and nothing happens, all this nonsense about acid rain will stop." Across the Atlantic, President Reagan's chief of air pollution control, Kathleen Bennett, cautioned that action to reduce acid rain would be "premature."[6] She explained, "Acid rain as a separate field of study is scarcely a decade old."

More pollution, taller smokestacks

As the modern industrial era blossomed, the demand for energy surged — and with it, the amount of pollution. In both Europe and the U.S., coal consumption skyrocketed into the mid-twentieth century, joined by a sharp rise in oil consumption after World War II. In a single generation, from 1950 to 1970, Europe's SO_2 emissions doubled and its NO_x emissions more than doubled.[7] In the United States, SO_2 emissions increased by 45% in the decade from 1960 to 1970.[8] Emissions dropped slightly after 1970, when concern about deteriorating air quality in the nation's cities led Congress to enact the Clean Air Act, but now SO_2 emissions have begun to rise again. NO_x emissions have more than tripled since 1940 and continue to rise.[9]

Taller and taller smokestacks were built in order to disperse emissions as far as possible from the source. ("The solution to pollution," one motto put it, "is dilution.") The height of the tallest smokestack in the United States doubled since the 1950s and the average height of smokestacks tripled.[10] Ironically, the Clean Air Act of 1970 encouraged the construction of ever-taller smokestacks: tall stacks enabled more pollution to be emitted without exceeding ground-level limits. Tall stacks thus became a substitute for pollution control. As a result, pollution was carried far beyond cities and industrial centers to the once-pristine countryside, and even further, beyond national boundaries. The whole of eastern North America and western Europe became the acid equivalent of the proverbial smoke-filled room. Acid rain was transformed from a local, urban problem into an international threat.

Glimmer of a problem

The first inkling that a large-scale problem was at hand came in the late 1950s and early 1960s, when Scandinavian scientists noted a disturbing connection between acid rain, the increasing acidity of lakes and streams, and the disappearance of fish. In 1968, the Swedish soil scientist Svente Odén summarized evidence that the acidity of precipitation in Scandinavia was increasing, and he showed from an analysis of the trajectories of air masses that much of the acid rain in Scandinavia originated with SO_2 emissions in England and central Europe.[11] Odén's work drew extensively on rain chemistry data from a network that, ironically, had been designed to study the beneficial effects on agriculture of nutrients deposited by precipitation.

The Stockholm newspaper, *Dagens Nyheter*, published an account of Odén's findings and sparked the first public awareness of the problem.[12] In 1972, the Swedish government brought its case before the U.N. Conference on the Human Environment with a document entitled, "Air Pollution Across National Boundaries: The Impact of Sulfur in Air and Precipitation." That same year, Norway launched a comprehensive eight-year study of the effects of acid rain on Norwegian forests and aquatic resources.[7]

The fears of Scandinavian scientists proved well-founded. Norway's seven southern rivers, which receive the brunt of acid rain, are now 30 times as acidic as the northern rivers (pH 5.12 vs. pH 6.57 average). In one decade, from 1966 to 1976, the acidity of these seven rivers doubled, and the salmon catch is now "near zero" in all of them.[13] Of the 700 small to medium-sized lakes in the southernmost counties of Norway, 40% are barren of fish. Another 40% have "sparse" populations. A survey of 5,000 lakes in the southern half of Norway found that trout populations had become extinct in 22% of the low-altitude lakes (below 660 ft.) and 68% of the high altitude lakes (above 2,620 ft.). In Sweden, a preliminary survey of more than 100 lakes in the south revealed that fish species such as trout, pike, minnow and roach have been lost in 14-43% of the lakes. It is estimated that several thousand lakes in Sweden are devoid of fish.

Across the Atlantic, Canadian and American scientists were fitting together pieces of a similar puzzle. In an important series of papers beginning in the mid-1950s, the ecologist Eville Gorham documented the effects of acid precipitation on Canadian lakes, vegetation and soils. Gorham was concerned with the effects of pollution from metal smelters, especially the giant nickel smelter at Sudbury, Ontario — the largest man-made source of SO_2 in the hemisphere. But Gorham's work went largely unnoticed by the scientific community. It was not until Svente Odén visited the U.S. and Canada in 1971 to deliver a series of lectures on acid rain that the scientific community sat up and took notice of the problem.

By the following year, the Canadian scientists Beamish and Harvey had documented the disappearance of fish from acidified lakes in the La Cloche Mountains in south central Ontario.[14] With the aid of historical records, they determined that the acidity of 11 lakes had increased 10- to 100-fold. "The increase in acidity," they concluded, "appears to result from acid fallout in rain and snow." Gene Likens at Cornell University warned that acid rain might be damaging lakes, soils, and forests in the northeastern U.S. In 1976, Carl Schofield reported a severe decline in fish in acidified lakes in the Adirondack Mountains.[15]

Typical of those acidified lakes is Lake Colden, whose 38 acres are nestled in the high peaks of New York's Essex County at an elevation of 2,764 ft. Records show that in the 1930s there were "excellent angling opportunities" for trout at Lake Colden.[16] By the late 1950s, however, "fisherman were beginning to register complaints" about the decline of trout fishing, despite an annual aerial stocking program. In 1965, the surface pH had dropped to 5.0, and by the 1970s the lake had become another two to three times more acidic. Surveys in 1970 and 1973 failed to find any trout whatsoever, and the stocking program was terminated.

Silver Lake in Hamilton County is another high-altitude jewel of a lake. In 1932 it had a healthy pH of 6.6 and was reported to have "good brook trout fishing despite heavy angling pressure."[16] By 1969, no fish life was detected, despite aerial stocking of more than 7,000 fingerlings yearly. In 1975, Silver Lake had a pH of 4.92, and a "final netting effort" failed to find a single fish.

A comprehensive survey of 1,000 Adirondack lakes and ponds revealed that nearly 25% have acidified and no longer support trout or other game fish.[16] Lakes and streams have acidified in sensitive areas in New England, New York, New Jersey, Pennsylvania, W. Virginia, and N. Carolina.[13] Fish kills are reported as far south as the Great Smoky Mountains of Tennessee, when melting spring snow sends a sudden pulse of acid into lakes and streams.[17]

Assessing the damage

Is acid rain a "catastrophe," as some have called it, or is it the wild concoction of "alarmists"? Everyone has seen cartoons showing an umbrella shredded by acid rain, but who has ever *seen* such an umbrella? Despite advertisements to "acid rain-proof" the family car, no one has ever seen a car disintegrate in a downpour. The real threat of acid rain is not its corrosiveness, but its much more subtle, cumulative effects. The evidence is now compelling that acid rain is damaging a wide variety of natural and man-made resources, including aquatic life, vegetation, soils, air quality, and man-made structures. Many of these effects are discussed in Chapter 5. A few of the major concerns are summarized here.

The death of fish is the most highly publicized effect of acid rain, yet even the loss of fish is due not just to the direct action of acidity itself but also to more subtle changes in the chemistry of lakes brought about by acid rain. It is now known, for example, that acid rain dissolves aluminum — one of the most abundant elements in the soil — and carries it into lakes and streams at levels toxic to fish. Aluminum interferes with salt balance, and produces a clogging of the gills that causes fish to literally suffocate to death.

The loss of fish is only the most visible sign of damage to a lake, and it is often the last sign. Many other aquatic creatures are even more sensitive to the effects of acid rain than fish. Acidification has led to the loss of freshwater clams, snails, crayfish, roundworms, tadpole shrimp, bottom-dwelling invertebrates, insects and other organisms that go unnoticed by humans but that are critical to the food chain and to the recycling of nutrients in a lake.[13] In the Northeast, salamanders hatching in acidic pools formed by melting spring snow have become deformed and died; and in Canada, the population of frogs and other animals that spawn in meltwater pools has declined. On a few acid lakes in Scandinavia, the number of birds such as goldeneye ducks has *increased*, because the birds feed on swarms of acid-tolerant insects that would normally have been kept in check by other predators. But around many acid lakes, bird populations have dwindled. In the Adirondacks, the cry of the loon is heard less frequently as populations have declined; mergansers, gulls, and kingfishers are particularly sensitive to the decline in fish and other aquatic life, and have been observed to feed only on lakes with pH greater than 5.6.[13] In short, acid rain is not merely a threat to individual species; it is altering and endangering entire ecosystems, in some cases irrevocably.

Recent evidence indicates that acid rain may be seriously damaging high-altitude forests in the eastern U.S., a finding that has added a new sense of urgency to the problem. On Camels Hump in northern Vermont, a site that receives highly acid rain and fog, the red spruce forest is now largely a graveyard of ghostly stumps[18] (see Plate I). More than half the red spruce have died since the 1960s, and a recent survey failed to find a single healthy spruce seedling.[19] Originally thought to be unique, Camels Hump is now considered an extreme example of a widespread phenomenon. Analysis of tree rings has revealed a severe decline in the growth of evergreens and some hardwoods, in areas ranging from New York's Adirondack Mountains,[20] New Hampshire's White Mountains, and New Jersey's Pine Barrens,[21] to as far south as Mt. Mitchell in North Carolina's Great Smoky Mountains. As other factors have been ruled out, one by one, the probable cause appears to be acid rain and related forms of air pollution, such as ozone.[22] (Ozone is toxic to vegetation and is also a respiratory irritant. Its formation in the lower atmosphere is enhanced by NO_x emissions. This ozone should not be confused with the beneficial "ozone layer" that is formed by natural processes high in the upper atmosphere and that absorbs much of the sun's harmful ultraviolet rays.) Damage to forests may well be the catalyst that spurs Congress to act on acid rain. In Germany — Europe's third largest emitter of SO_2 — severe damage to forests has led that nation to reverse its long-standing "wait and see" policy on air pollution and to initiate a major program to reduce emissions (see Chapter 5).

Acid rain poses an indirect threat to human health. Lead, mercury and other toxic metals in the environment are more soluble in acid waters; these metals are being leached from the environment and carried into drinking water supplies and into the food chain. A comprehensive survey of drinking water in New York and New England found that in 8% of households surveyed, water samples allowed to stand overnight in household plumbing contained lead levels exceeding the E.P.A.'s Maximum Contaminant Levels.[23] More than 40% of the water samples exceeded the Secondary Maximum Contaminant Levels for copper. Indeed, elevated lead levels have been found in the blood of children in the Adirondacks whose drinking water is supplied from acidified wells.[24] Toxic metals have been accumulating in the food chain as well. Fish from some acidified lakes in New York, Canada, and Scandinavia now contain mercury at levels toxic to humans.[25] To date, there have been no reports of acute toxic metal poisoning attributable to acid rain. However, lead and mercury can accumulate in the body, and even small amounts of these metals can damage the nervous system, especially in children. The burden of toxic metals in drinking water may pose a significant public health problem even in the absence of dramatic manifestations.

Acid rain is so widespread in eastern North America, and touches so much of the environment, that few scientists are confident they can predict the full range of future effects of acid rain. In fact, serious effects may even go undetected until it is too late to reverse them. The following three concerns illustrate the very wide variety of *potential* effects of acid rain.

• Some pesticides and herbicides are degraded more rapidly in acid waters and might have to be applied more often or at higher doses. Others, such as malathion and parathion, are degraded more slowly at low pH levels, and may therefore persist in the environment. Some pesticides that are normally negatively charged can become electrically neutral at lower pH; this can enhance the rate at which they are absorbed by plants and animals, and can increase their toxicity.[13]

• Acid deposition in the Arctic may slow the growth of lichens, which are the primary food source for the great caribou herds. Lichens are particularly sensitive to sulfur pollution, and their growth is already so slow that any additional perturbation may wipe out the caribou's food source.

• Acid rain may disrupt some of nature's microbial eco-systems, many of which are now poorly understood. For example, several insect species are in part kept in check by "good" viruses that are found on leaves and elsewhere in the environment. If the protein coat of these viruses were disrupted by acid rain and the viruses were rendered less infective, it could lead to an explosion in some insect populations.

It should be emphasized that these concerns are entirely speculative at present; nevertheless, the concerns are based on plausible scenarios.

Four general considerations argue that acid rain is a serious and even urgent problem that needs to be addressed without delay. First, the number of resources threatened by acid rain is large. About one quarter of the surface water in the eastern U.S. — some 14,000 square miles — is moderately to highly sensitive to acid rain.[26] In Canada, nearly 11,000 square miles of surface water are highly sensitive to acid rain. According to U.S. and Canadian estimates, the number of acidified lakes in eastern North America will double by 1990 at current rates of acid deposition. In addition, nearly 75% of the forest area in the eastern U.S. now receives at least moderately acidic deposition, and 10% receives highly acidic deposition. The areas most sensitive to acidity include wilderness and recreation areas which are already at a premium in the eastern U.S.

Second, the extreme complexity of most ecosystems has made scientists wary of putting an upper limit on the future toll of acid rain. Few scientists believe that large-scale ecological catastrophe is imminent, but the localized "mini-catastrophes" in the Adirondacks and on Camels Hump are a stark warning that large-scale ecological collapse is indeed possible.

Third, many of the effects of acid rain are irreversible on human time scales. While fish have been restored to some acidified lakes to which limestone has been added, the most severely affected lakes will not be restored to their pristine states. The destruction of evergreen forests on Camels Hump and elsewhere, and the degradation of soils and of whole ecosystems, will not be reversed in our lifetimes.

Finally, a long lead time is required to reduce acid rain — as much as a decade to install the massive pollution-control equipment needed to reduce SO_2 emissions by 50%. Since our ability to respond to acid rain will always lag several years behind our knowledge of the dangers, it is prudent to take preventive action at the earliest warning signs. Maintaining an adequate margin of safety has long been a cornerstone of public health policy and water resource policy.

What can be done to reduce acid rain?

One way to reduce acid rain in the long-term is to burn less fossil fuel — by conserving energy and by developing alternative sources of energy such as solar, wind, hydro and nuclear power, none of which contribute to acid rain. However, even under the most optimistic projections, these alternative energy sources are not likely to make major inroads until at least the end of this century.

The one fuel that is certain to play an increasing role is coal. It is cheap and abundant: the nation's coal reserves are sufficient for several centuries at the current level of energy demand. In the wake of the 1973 Arab oil embargo, power plants have been under economic pressure to convert from oil to coal. Before the embargo, 44% of the nation's electricity came from coal; now 53% does. It is unlikely that acid rain will be significantly reduced without imposing strict controls on SO_2 and NO_x emissions from the burning of coal.

There are three major ways to reduce SO_2 emissions from coal combustion: 1) burn low-sulfur coal (less than 1% sulfur by weight) rather than high-sulfur coal (greater than 3% sulfur); 2) remove the sulfur from coal before it is burned ("coal cleaning"); and 3) use pollution-control devices to remove the sulfur during or after coal combustion. In a large-scale program to control acid rain, say one that required a 50% reduction in SO_2 emissions, all three of these methods would come into play.

Opposition from the energy industry

Legislation proposed by a number of U.S. senators from the Northeast would place a "cap" on SO_2 emissions from power plants — that is, a limit on the amount of SO_2 emitted per unit of electricity generated. Under such a plan, the majority of emissions reductions would come from the "dirtiest" coal-fired plants, chiefly clustered in the Midwest. Legislation aimed at reducing SO_2 and NO_x emissions has been adamantly opposed by the midwestern electric utilities and the coal industry. To understand the depth of the industry's opposition, it is important to understand what a large-scale pollution-control program would entail.

Switching from high-sulfur to low-sulfur coal is viewed with alarm in the Midwest, because it would disrupt the high-sulfur coal industry. Most of the coal burned by midwestern power plants is high-sulfur coal, mined in Ohio, Illinois, Indiana, Pennsylvania, western Kentucky, and northern West Virginia. Coal production in this region would decrease by 10 to 20% for a major acid rain control program.[27] "That land is already so poor," pleaded one congressman from a coal-mining district, "that even the jack-rabbits carry knapsacks." According to Larry Reynolds of the United Mine Workers, reducing SO_2 emissions by 40% would throw 83,000 high-sulfur coal miners out of work.[28] However, that estimate assumes that virtually all the emissions reductions would come from switching to low-sulfur coal, rather than from pursuing a mix of options, including installation of pollution-control devices. The Congressional Office of Technology Assessment (O.T.A.), a non-partisan agency that researches issues at the request of members of Congress, estimates that 9,000 to 38,000 jobs would be lost in the high-sulfur coal mining industry during the decade of implementation.[27] (For comparison, the coal mining industry employed 254,000 people nationwide in 1982.)

On the other hand, production and employment would *increase* in the areas that produce low-sulfur coal: central Appalachia (southern West Virginia, eastern Kentucky, Tennessee, Virginia) and western states such as Wyoming, Montana, and Idaho. Western low-sulfur coal is plentiful but expensive; the cost of shipping it east by railroad is half the total cost of the coal. Efforts are underway to develop a "coal-slurry pipeline" (in which coal is crushed, suspended in water, and forced through a pipe much like oil), but the prospects for a pipeline are not bright.[29] The railroads have refused to grant the needed rights-of-way on their land, and the Reagan Administration has been reluctant to acquire the land by eminent domain, preferring to leave the matter to individual states. A pipeline would also require large quantities of scarce western water.

Some observers contend that rising productivity, not acid rain, is the greatest threat to employment in the high-sulfur coal industry. Nevertheless, it appears that the politically expedient solution to acid rain may be one that relies heavily on expensive pollution-control devices which allow high-sulfur coal to be burned. In any case, coal-switching is not expected to yield more than one-third the emissions reductions envisioned in a large-scale control program. Furthermore, the Department of Energy estimates that "coal cleaning" (see Chapter 6) can reduce SO_2 emissions below current levels by no more than about 2 million tons per year — a 10% reduction.[30]

Therefore the bulk of emissions reductions must come from pollution-control devices. These include "scrubbers" and more exotic technologies currently under development (see Chapter 6). Scrubbers can remove up to 90% of the SO_2 from coal combustion, but they carry a price tag of at least 100 million dollars each.

The utilities have opposed legislation that would require "retrofitting" scrubbers onto existing power plants. They point out that the current Clean Air Act requires that *new* power plants be equipped with the latest pollution-control devices to remove 80% of the SO_2 produced. (These are the so-called "new source performance standards" of the Clean Air Act.) According to this view, existing legislation and the passage of time would solve the acid rain problem. However, boilers last a long time — 35 to 50 years or more — and there is no assurance that existing boilers would be retired even by the end of the century. On the contrary, the financial incentives are strong for keeping old power plants running as long as possible, in order to avoid having to install expensive control technology on new plants. Also, energy conservation has reduced the demand for electric power, and with it the demand for new power plants. In order to significantly reduce SO_2 emissions within a reasonable time, therefore, it will be necessary to retrofit existing plants with pollution-control devices.

How much will it cost?

The annual cost of reducing SO_2 emissions by 40% has been estimated by the utilities and by others to be in the range of 1.9 to 5.8 billion dollars per year.[27] The upper limit is considered by many observers to be high: it assumes that the reductions come entirely from scrubbers with no coal-switching, and it includes the cost of offsetting projected increases in SO_2 emissions, estimated to be an additional 4 million tons by 1990.

The electric power industry has claimed[31] that if these costs were passed along to the consumer, electric rates in the Midwest would rise from 18% to as much as 63%. But this estimate too is for the "worst case" and is considered high. The Office of Technology Assessment puts the increase at no more than 12 to 16% for any state[27]; for Ohio, Illinois, West Virginia and Tennessee, the increase would be 6 to 9%.

The British electric power industry also appears to have initially overestimated the costs of reducing SO_2 emissions.[32] It recently cut its cost estimates in half in the face of mounting pressure from the European Economic Community for a 60% reduction in SO_2 emissions in Europe by 1995. Britain's Central Electricity Generating Board had reported in the fall of 1983 that abatement would cost 4,000 million pounds; in the spring of 1984 the Board revised its estimate downward to 1,400 million pounds, or 120 million pounds per plant. The price of electricity to consumers would increase by 4% according to current industry estimates.

The cost of cleaning up acid rain has been described by the energy industry as "gigantic" and "exorbitant", but these epithets do not put the cost in perspective. Just the cost of hauling coal by railroad now adds five billion dollars annually to America's electric bill,[29] and this cost has skyrocketed since the railroads were deregulated in 1980. (A federal lawsuit brought by Western Fuels, a coal cooperative, charges that the Burlington Northern Railroad alone will overcharge coal companies up to 8 billion dollars over the next eight years.) The electric power industry itself has capital assets of more than 1 trillion dollars. The cost of reducing emissions is thus in keeping with the size of the industry.

Who should pay for cleaning up acid rain?

Congress has been considering a variety of plans for financing the clean-up of acid rain that would shift the cost from the utilities and coal companies to a broader base. One proposal would impose a tax on all industrial SO_2 emitters; coal-fired sources would pay a higher tax than oil-fired sources because of their "dirtier" emissions. Another proposal would impose a tax on all power plants based on the amount of electricity they generate, regardless of the amount of pollution they emit. Still another proposal would tax all fossil fuels, including gasoline. The proposal that distributes the burden most widely is to finance the clean-up with a "general budgetary outlay" − that is, let the taxpayers foot the bill. The northeastern states believe the fairest approach is to "let the polluter pay," but some observers believe the politically expedient solution to acid rain will be to distribute the financial burden of clean-up as widely as possible.

What about reducing NO_x emissions?

Is it fair that Congress has focused on reducing SO_2 emissions, even though NO_x emissions also contribute to acid rain? There are several reasons for this selective attention. In the eastern U.S., more than two thirds of the acidity in rain comes from SO_2 emissions, while only one-third comes from NO_x emissions. Furthermore, the oxidation of SO_2 produces *twice* as much acidity as the oxidation of NO_x (see Chapter 3). Therefore, molecule for molecule, it makes more sense to reduce SO_2 emissions. Acid sulfate appears to play a somewhat greater role than acid nitrate in acidifying lakes and streams, because acid sulfate runs off more easily into lakes and streams while acid nitrate is retained more readily by the soil. Also, SO_2 emissions are usually carried further downwind than NO_x emissions, and contribute more to interstate and international air pollution. The technologies for reducing SO_2 emissions are currently more numerous, more practical and less expensive than those for controlling NO_x. Finally, it is easier to monitor the level of SO_2 emissions than NO_x emissions, an important consideration for a regulatory agency.

However, there is growing concern that increases in NO_x emissions might offset any benefits from reducing SO_2. Nitrogen oxide emissions have more than doubled in the past 25 years, and are projected to increase through the end of the century.[19] As a result, the level of acid nitrate in rain is on the rise in the Northeast, the Southeast,[33] the Rockies,[34] and Southern California. Acid nitrate is relatively more abundant in winter than in summer, and is in part responsible for the "acid shock" to lakes that occurs in spring when acid snow melts and sends a sudden pulse of acid into a lake.[35] NO_x emissions are also thought to promote the formation of ozone and other chemical oxidants in the air that speed up the formation of acid rain and are toxic to plants in their own right.[9] For all these reasons, it is considered important to limit or reduce NO_x emissions as well as SO_2.

One reason for the sharp increase in NO_x emissions during the past few decades is that industrial boilers have been operated at ever-higher temperatures in order to increase their efficiency. The higher the temperature of a furnace, the more nitrogen and oxygen in the air combine with each other, and the more NO_x produced. Some of the new combustion technologies allow fuel to be burned efficiently at *lower* temperatures, producing less NO_x (see Chapter 6). In a Senate bill originally proposed by Sen. George Mitchell (D., Maine), utilities were given the option of reducing NO_x emissions in lieu of an equivalent reduction of SO_2 emissions. Motor vehicles have also been eyed as possible targets for control, since they account for nearly half of all NO_x emissions.

Environmental trade-offs

Virtually all of the methods for cleaning up acid rain create other environmental problems.[36] Switching to low-sulfur coal will accelerate strip-mining in the West. Installing "scrubbers" and other pollution-control devices will aggravate the problem of how to dispose of millions of tons of sulfur-containing sludge. (In some control technologies, sulfur is removed in the form of gypsum — calcium sulfate — and can be sold as a useful product. In Canada, sulfur is removed from natural gas and used to manufacture sulfuric acid for industrial use.[37] Unfortunately, the market for sulfur products is not large enough to accommodate the huge quantities of sludge generated by scrubbers. No matter how ingenious the recovery process, therefore, waste disposal will always be a concern.) Even the new technologies for converting coal into a liquid or a gas produce toxic byproducts that are hazardous both to plant workers and to the environment. Reducing NO_x emissions from motor vehicles entails a trade-off in reduced gas mileage, increasing the nation's dependency on oil. Alternatives to fossil fuel combustion — such as nuclear power — pose their own well-publicized hazards.

In short, cleaning up acid rain will be costly — economically, socially, politically, and even environmentally. Nevertheless, the evidence is clear that acid rain must be confronted as a serious problem that will not vanish of its own accord.

Science and policy diverge

At a time when science and policy ought to be converging on solutions to acid rain, they have diverged. Every major science advisory panel studying the problem has recommended substantial reductions in SO_2 and NO_x emissions in order to reduce the acidity of rain. Yet the opponents of acid rain controls — the midwestern utilities, the coal industry, and their supporters in the Reagan Administration and Congress — have called only for "more research."

In June 1981 the National Academy of Sciences released a comprehensive report on the effects of acid rain and other consequences of fossil fuel combustion.[38] Citing "the probability of a crisis in the biosphere," the report concluded that "continued emissions of sulfur and nitrogen oxides at current or accelerated rates, in the face of clear evidence of serious hazard to human health and to the biosphere, will be extremely risky from a long-term economic standpoint as well as from the standpoint of biosphere protection."

The Academy's report called for a 50% reduction in the acidity of rain in the Northeast. (The recommendation was based on the simple observation that sensitive regions receiving rain with average pH greater than 4.6 to 4.7 have not yet suffered adverse effects. Rainfall in the Northeast is still more than twice as acidic as this guideline. See Chapter 6.) "Of the options presently available," the Committee concluded, "only the control of emissions of sulfur and nitrogen oxides can significantly reduce the rate of deterioration of sensitive freshwater ecosystems." The report further stated that "control of SO_2 from new electrical generating plants alone would be insufficient to accomplish this [reduction in acidity], and thus restrictions on older plants must be considered."

The Academy's report was dismissed by the Reagan Administration as "lacking in objectivity," though the Administration did not take issue with the scientific findings. In October 1981 the Environmental Protection Agency issued a press release reiterating the Administration's position that "scientific uncertainties in the causes and effects of acid rain demand that we proceed cautiously and avoid premature action."[6] To reduce emissions further, the E.P.A. warned, "would increase the tensions already evident among states."

The E.P.A. released its own "Critical Assessment" of acid rain in 1982 — a long-awaited 1200-page document that was the combined effort of 54 scientists from universities and research institutes around the country.[13] Yet E.P.A. administrators appeared not to have read the agency's own report. According to the journal *Chemical and Engineering News*, "scientists involved with the report concede that even a cursory review of the chapter summaries can make a case for the linkage between man-made emissions in the Midwest to dead or dying lakes, aquatic life and forests, and to materials damage, and to the potential harm to human health in the eastern U.S. and southeastern Canada." But the E.P.A.'s chief of air pollution control, Kathleen Bennett, downplayed the report, arguing, "Any attempt to draw integrated conclusions from the various chapters at this point is scientifically unsupportable."

Notably missing from the E.P.A. report were even tentative answers to several politically-charged but key aspects of the problem: Which sources of SO_2 and NO_x are responsible for acid rain in the Northeast and other vulnerable regions? How many more lakes would acidify in the future if action on acid rain were delayed? In one section of the report, an in-house E.P.A. official cited three models for predicting when a lake will acidify, but then declined to use the models, calling it "a very long way to go before any model can be used with quantitative confidence."[39] The official concluded that "none of the three models discussed briefly here have been verified adequately for 'off-the-shelf' application in North American waters. Such an application . . . would violate virtually every rule concerning the prudent use of predictive models." Thus readers of the report were left without even a crude estimate of how many more lakes might be at risk to acid rain.

Another major report[26] on acid rain was released in February 1983: a bilateral study by the U.S. and Canada that was initiated by President Carter in 1980 when the two nations signed a Memorandum of Intent to study the causes and effects of acid rain and to develop a mutual agreement on limiting transboundary air pollution. The study had been intended to reach a consensus on the science, but the U.S. and Canada deadlocked over a key section dealing with the effects and significance of acid rain. In an unusual move, the two nations issued separately worded conclusions. The Canadians concluded that reducing acid rain "would reduce further damage" to sensitive lakes and streams. The U.S. version omitted the word "damage" and substituted "chemical and biological alterations." The Canadians concluded that "loss of genetic stock would not be reversible." The U.S. version omitted the sentence altogether. The Canadians proposed reducing the amount of acid sulfate deposited by precipitation to less than 20 kilograms acid sulfate per hectare per year — roughly a 50% reduction — in order to protect "all but the most sensitive aquatic ecosystems in Canada." The U.S. version declined to recommend any reductions.

"We committed ourselves," protested John Roberts, Canada's Minister of the Environment, "to reduce our sulphur dioxide emissions east of the Saskatchewan/Manitoba border by 50 percent by 1990 contingent on parallel action in the United States. This represented a doubling of the cutbacks that had already been initiated unilaterally in Canada . . . . Based on the conclusion of the Final Reports of the Work Groups, I urge the U.S. to reconsider the proposal we have made."

Originally, the U.S./Canada report was to have been peer-reviewed by the National Academy of Sciences, but the Administration balked, preferring to keep the review process under the aegis of the White House. Congressional proponents of acid rain legislation were livid. Senator George Mitchell interrupted Senate hearings on acid rain to protest. "I am deeply distressed by a report which was just handed to me," he said, "which indicates that the Reagan Administration has refused a National Academy of Sciences request to conduct a scientific review of U.S.-Canadian studies on acid rain and instead has assigned a White House panel to carry out the study. Sources within the Academy and the Environmental Protection Agency said the White House refused to permit the Academy to perform the peer review study because the Academy has previously issued a report calling for additional pollution control to deal with the acid rain phenomenon." A former White House environmental advisor, James McAvoy, confirmed that "As a result of that (Academy) report, we were very concerned about their objectivity."[40]

The scientific panel that the White House handpicked to carry out the review sent the President a report that again strongly recommended action on acid rain. Under the chairmanship of William Nierenberg, Director of the Scripps Institute of Oceanography, the panel concluded that, "It is in the nature of the acid deposition problem that actions have to be taken despite incomplete knowledge . . . If we take the conservative point of view that we must wait until the scientific knowledge is definitive, the accumulated deposition and damaged environment may reach the point of 'irreversibility' . . ."[41]

The panel considered the acid rain problem so pressing that it delivered its five-page interim conclusion to the White House by June 1983, well in advance of the full report, which was completed in May 1984. By August 1984, the White House had still not released the full report, leading some members of Congress to allege that the report had been "suppressed." "Let me say that somebody in the White House ought to print the damn thing," said Nierenberg. "I'm sick and tired of it."[42]

Politics, but no policy

A policy on acid rain appeared in the offing when William Ruckelshaus assumed leadership of the beleaguered Environmental Protection Agency in 1983. (The Agency's former chief, Anne McGill Burford, had resigned amid charges of mismanagement and conflict of interest. Mrs. Burford was later appointed chairman of the National Advisory Committee on Oceans and Atmosphere — a group which also advises the President on acid rain — but she again resigned under fire after calling the position a "nothingburger" and the Committee "a joke".) Ruckelshaus described acid rain as one of the "cosmic issues" confronting the nation, and as "the most difficult, complex public policy issue I have ever faced." He made it one of his top priorities to present "four or five options" on acid rain to President Reagan no later than the end of September.[43] A study group appointed by Ruckelshaus drafted a proposal that included a 50% reduction in SO_2 emissions in 21 eastern states — an option that the group said would provide "the most uniform reduction" in acid rain in the Northeast and Canada.[44] But in mid-October, Mr. Ruckelshaus announced that the Administration's policy on acid rain had been "delayed."[45] By the end of October a policy on acid rain was postponed "indefinitely." Ruckelhaus was under intense pressure from the White House Office of Management and Budget (O.M.B.) to go slow on acid rain. The budget office went so far as to delete several key passages from an E.P.A. report entitled, "Environmental Progress and Challenges: an E.P.A. Perspective." The E.P.A.'s final draft had argued that "acid deposition is believed to pose little direct risk to human health," but conceded: "However, the National Academy of Sciences reported that acid deposition is a threat to human welfare because of its potential impact on materials, forest and farm productivity, aquatic ecosystems and drinking water systems." That sentence was simply crossed out by the O.M.B. The final draft had said, "E.P.A. will be working with the states to prepare to implement an acid rain control program." The budget office reworded the passage, calling for more research to "increase our ability to determine the need for and select effective controls."[46]

Behind the scenes, the E.P.A. was seeking changes that would have *increased* acid rain. For example, in November 1983, the E.P.A. proposed to change the way in which coal-fired plants monitor their SO_2 emissions — a change that would allow an estimated 700,000 tons more SO_2 annually.[47] The E.P.A. also tried to relax regulations governing the use of tall smokestacks, but was rebuffed by the U.S. Court of Appeals, which ruled 3-0 in October 1983, that the Agency's actions were "arbitrary and capricious" and "violated the strict interpretation (of the Clean

Air Act) that Congress specifically intended."[48] The E.P.A. also sought to allow a doubling in NO_x emissions from automobiles.

The Mobil Corporation ran an advertisement on the editorial pages of several newspapers in October 1983, lauding a speech Mr. Ruckelshaus had given before members of the National Academy of Sciences. "It's refreshing indeed," wrote Mobil, "to hear the nation's number one pollution policeman ask that Americans 'reject the emotionalism' that mires any discussion of environmental quality and paralyzes 'honest public policy' . . . Too often, U.S. environmental policy has been based on political considerations . . ."

Congress fails to act

Public support for cleaning up acid rain extended far beyond the groups who might normally be termed "environmentalists." One industry journal, *Chemical Processing*, polled its readers on acid rain in 1984 and found that 72% of the more than 1,000 respondents believed that "the cost of retrofitting coal-burning facilities with scrubbers" would not "outweigh the possible environmental benefits."[49] The same percentage believed that "the environmental benefits accrued by cutting back sulfur oxide emissions would be worth the consequent loss of jobs." Eighty-five percent believed that "neutralizing damaged lakes with limestone would be effective only as a temporary measure." There was less agreement on who should pay for the cleanup. Sixty-two percent agreed that "the most polluting states should be taxed most heavily," while 38% believed the cost of emission controls should be "shared equally by electricity generators across the country."

Despite strong public sentiment, legislation to control acid rain has been stalemated in Congress. Acid rain is virtually unique among environmental problems in dividing Congress along regional rather than party lines. For example, two members of the Senate who have been most active in introducing legislation to clean up acid rain have been the chairman of the Senate Environment Committee, Robert Stafford (a Vermont Republican), and George Mitchell (a Democrat from Maine). By contrast, the Senate Energy Committee has strongly backed the position of the energy industry, led by such senators from coal-mining states as Wendell Ford (D., Kentucky) and Richard Lugar (R., Indiana), who termed acid rain controls "extravagant and odious to the Midwest."

The real bottleneck has been in the House of Representatives. A major bill for cleaning up acid rain was proposed by the chairman of the House Subcommittee on Health and Environment, Henry Waxman, a liberal Democrat from California who says reluctantly, "I guess I'm an activist." But the Environment Subcommittee is heavily weighted with representatives from Ohio, a state which generates 96% of its electric power from coal and which would bear the brunt of acid rain controls. The Subcommittee's ranking member, Dan Madigan, represents a coal-mining district in Ohio and is less than eager to see an acid rain control program. "Isn't it true," asked the congressman of one Canadian witness, "that acid rain is a conspiracy cooked up by the Canadian government in order to sell more electric power to the United States?" (According to the "conspiracy" theory, controlling acid rain would raise the cost of generating electricity in the Midwest to the point where utilities would find it cheaper to purchase electricity from the Canadian grid — a windfall for Canada.) Ironically, the deciding vote that killed acid rain legislation in the House Environment Subcommittee in 1984 was cast by a liberal, environmentalist Democrat: Dennis Eckhart from Ohio.

As the 1984 presidential campaign got underway, the coal industry took the offensive. "Acid rain is only a facade for the forces who would deliberately destroy the carefully-crafted balance of environmental and economic goals that have been achieved by the Clean Air Act during the last decade," said Carl

E. Bagge, president of the National Coal Association.[50] Accusing the Democrats of "pandering to the environmentalists and to the public opinion polls," Bagge told the American Power Conference, "I ask you, how can anyone compare them with our President, who has refused to be a party to the fraud? I ask you, on the basis of this sharp contrast, who do you believe should have access to the levers of political power in this nation?" The latter question is a profound one. Certainly the coal industry was likely to have "access to the levers of political power." Consider that the nation's largest coal company (Peabody Coal) is wholly owned by the giant multinational firm (Bechtel) whose former president and director (George Shultz) left to become President Reagan's Secretary of State. Indeed, it was Mr. Shultz who informed the Canadians in October 1983 that a policy on acid rain would not be forthcoming from the United States.

A philosophy emerges

One of the most intriguing questions at the core of acid rain policy is, How bad must the damage be before action is warranted? This is ultimately a question of values, not science. To some, a single dying lake would warrant a billion-dollar control program. At the other extreme are those who "would carry the landscape to market", in Thoreau's words, "if they could get a good price for it." The Reagan Administration avoided making explicit its values: the "uncertainty" that it touted was not so much in the effects of acid rain — which had been documented in hundreds of scientific papers — the uncertainty was in where the Administration had set the trigger for action.

Instead, members of the Administration put forward a remarkable philosophy that is so radical a departure from the norms of policy-making that it is worth examining. That is the notion that in order to act on acid rain, the "dollar cost of damages" must be reckoned greater than the cost of a control program. Acid rain became a problem of numbers, of balancing a kind of imaginary ledger. The philosophy surfaced repeatedly in Administration testimony ("Economic damage to lakes is expected to be small . . ."), but it is best crystallized in the remark of one of the staunch opponents of acid rain controls, David Stockman, Director of the Office of Management and Budget. He asked, "How much are the fish worth in these [Adirondack] lakes? . . . Does it make sense to spend billions of dollars controlling emissions from sources in Ohio and elsewhere if you're talking about very marginal volume of dollar value . . . ?"[51]

Stockman's comment embodies a philosophy that not only demeans our natural resources, which after all never came with price tags, but also corrupts a powerful and useful economic tool. Cost-benefit analysis was originally intended to enable policymakers to choose among alternative plans of action, to compare the costs and benefits of one plan with those of alternatives. It was never intended that the "dollar value" of benefits (even assuming they could be measured) must exceed the dollar cost of action, that action should be determined by numbers rather than human values. The reason is that most of the benefits we seek in public and private policy are not adequately expressed in dollars. (For example, when we invite friends for dinner, we may debate whether to serve lobster or hamburger, taking into account the cost of different menus and the state of our pocketbook. But we would never dream of putting a dollar value on our guests' friendship.) Nevertheless, in an effort to accommodate this philosophy, the northeastern states scrambled to make elaborate estimates of the economic losses from lost fishing licenses, decline in tourism, and the like — a desperate effort to make the "cost" of damages exceed the cost of cleanup. In an attempt to put dollar signs on the environment, some economists have resorted to "willingness to pay" surveys ("How much would you pay for clean air?") or have inferred such numbers from, say, the difference in real estate values in communities with polluted air versus communities with clean air. In part using these methods, economic losses due to acid rain and other forms of air pollution have been reckoned at several billion dollars a year.

In reality, virtually no public policy is predicated on Mr. Stockman's brand of cost-benefit analysis, let alone on "willingness to pay." Social security, defense spending, welfare, the Vietnam War, the space program — whatever their merits, any of the major policy decisions of the last few decades illustrate that there are things we do collectively as a nation simply because we believe they should be done, even though no one of us would do them individually. We have preferred to let our human values guide our public policy, even in cases where that has strained our pocketbooks.

Going sour

No one can seriously doubt that we are strong enough and clever enough as a nation to burn coal cleanly, to protect the jobs of miners, to protect the economic interests of one region as well as the environmental interests of another, and still preserve that elusive ideal of fairness. But if we are to redirect a dialogue that is going all too sour, it will require a cooperation and earnestness that we have not yet tapped. It will require going beyond the lament of one industrialist, "It's *our* cost and *their* benefit!" It will mean recognizing that national problems call for national solutions. No one has put it quite so aptly as Senator Mitchell:

"The waters, the resources of New England don't just belong to the people of New England. We don't protect the Grand Canyon just for the people of Arizona. We don't preserve the redwood forest just for the people of California. And the Grand Teton Mountains don't just belong to the people of Wyoming. There are certain resources that are natural and national in scope, and they belong to all of the people, and they deserve protection."

PART 2

THE SCIENCE OF ACID RAIN

2

THE COMPOSITION OF RAIN

Rain holds a mirror to the environment. From the chemical content of rain we can infer much about the quality of the air in which it formed. This chapter compares the composition of rain from the U.S. with that from various sites around the world. Rain in the eastern U.S. is found to contain an average of 10 to 30 times as much acid sulfate and acid nitrate as rain from pristine regions.

What's in rain?

Even unpolluted rain can contain both acids and bases. For example, carbon dioxide naturally present in air dissolves in rain to form carbonic acid (H_2CO_3), the weak acid responsible for the fizz in soda pop. If there were no other substances in rain, CO_2 would lower the acidity from neutral (pH 7.0) to weakly acidic (pH 5.6). Unpolluted rain also contains small amounts of acid sulfate and acid nitrate that are produced in the stratosphere. These acids would further lower the natural pH of rain to about 5.4. Several alkaline substances partly neutralize the acidity in rain: *Ammonia*, which is naturally present in the air, dissolves in rain to form ammonium salts. *Soil dust* containing alkaline minerals such as limestone (calcium carbonate) and dolomite (magnesium carbonate) can dissolve in rain and raise its pH. *Fly ash* from coal combustion contains calcium and magnesium oxides, which also are alkaline. The pH of rain therefore reflects the competing influence of several different substances.

By far the major constituents of rain in eastern North America are acid sulfate and acid nitrate.[52] As we will see in the next chapter, these acids come overwhelmingly from man-made pollution. Table 1 shows the average content of the major ions in rain from five different regions of the country: upstate New York, Ohio, northern Wisconsin, S. Carolina, and Oregon. Rainfall in upstate New York contains twice as much acidity, sulfate, and nitrate as rain from S. Carolina, and nearly ten times as much of these pollutants as rain from Oregon. Note that the overall acidity of rain (the concentration of hydrogen ions) is roughly equal to the sum of the ions derived from acids (sulfate plus nitrate) minus the ions derived from bases (ammonium, calcium, and magnesium).

Distribution of Acid Rain

Eastern N. America and W. Europe: The most acidic precipitation in the world falls on eastern North America and western Europe. Precipitation in these areas averages 60 to 100 microequivalents acid per liter, corresponding to pH 4.0 to 4.3. This is 10 to 30 times as acidic as unpolluted precipitation; individual storms can be several hundred to more than a thousand times as acidic as unpolluted precipitation. On average, two thirds of the acidity is due to acid sulfate, and one-third is due to acid nitrate. The average concentrations of sulfate and nitrate in precipitation in the eastern U.S. were 51 and 27 microequivalents/liter, respectively, for 1980.[53] The geographic distribution of acidity, sulfate and nitrate in North American precipitation[54] is shown in Figures 1, 2 and 3. Weak organic acids and small amounts of hydrochloric acid (HCl) are found in precipitation but do not contribute significantly to its acidity.[55]

Western U.S.: West of the Mississippi, precipitation is generally neutral or alkaline, due to suspended soil dust containing alkaline carbonates. In the Colorado Rockies, however, precipitation is often acidic. Acid sulfate and acid nitrate contribute equally, but the level of acid nitrate is on the rise.[34] In Pasadena[56] and other parts of Southern California, rain is as acidic as pH 3.9, and invididual fogs have reached as low as pH 1.7. Two-thirds or more of this acidity is acid nitrate, caused by NO_x emissions from automobiles.

Eastern Hemisphere: In Japan,[57] the pH of rain in industrial areas is often between 3.0 and 4.0. In urban areas in China, rain is as acidic as pH 2.25, due primarily to nitric acid, apparently from the burning of local coal which is low in sulfur but high in nitrogen.[58]

Remote Areas: In Antarctica[59] and Greenland[60], the average concentration of sulfate in precipitation was less than 2 microequivalents/liter. In Poker Flat, Alaska; Venezuela; Samoa; Katherine, Australia; and Amsterdam I., S. Indian Ocean, the average concentration of sulfate in precipitation was 7.1, 2.7, 1.5, 5.5, and 8.8 microequivalents/liter, respectively; and the average nitrate concentration was 1.9, 2.6, 0.2, 4.3, and 1.7 microequivalents/liter.[61] Compare these numbers with Table 1. Moderately acidic rain in some remote areas of the globe may be due to local biological sources of sulfur, or to long-range transport of man-made pollutants (see Chapter 4).

In summary, precipitation in the eastern U.S. contains some of the highest levels of acid in the world — 10 to 30 times the "baseline" level of acid sulfate and acid nitrate from pristine regions.

3

THE CAUSES OF ACID RAIN

Acid rain in the eastern U.S. is caused almost entirely by man-made emissions of sulfur and nitrogen oxides from the burning of fossil fuels. This conclusion is based on two lines of evidence: an understanding of the chemical reactions that sulfur and nitrogen oxides undergo in the atmosphere; and a quantitative tally, or "budget," of how much sulfur and nitrogen is emitted from various sources and how much returns to earth.

Chemistry of acid formation

The overall chemistry leading to the formation of acid rain has been known since at least the middle of the 19th century.[5] When a fuel is burned, sulfur and nitrogen in the fuel combine with oxygen in the air to form sulfur and nitrogen oxides. (The sulfur oxides are chiefly sulfur dioxide, SO_2, and lesser amounts of sulfur trioxide, SO_3. The nitrogen oxides are a mixture of nitric oxide, NO, and nitrogen dioxide, NO_2. The nitrogen oxides are abbreviated NO_x, where "x" stands for 1 or 2.) In contact with air, SO_2 and NO_x are completely oxidized to form acid sulfate and acid nitrate, respectively. The overall reactions consume oxygen and water:

$$SO_2 + \tfrac{1}{2} O_2 + H_2O \rightarrow 2H^+ + SO_4^=$$

$$NO + NO_2 + O_2 + H_2O \rightarrow 2H^+ + 2NO_3^-$$

These reactions take place spontaneously, and are "driven" by the fact that the end-products, sulfate and nitrate, are chemically very stable. (The world's oceans, for example, are filled with sulfate and nitrate of ancient origin.) From the chemist's viewpoint, acidity is a *byproduct* of the complete oxidation of sulfur and nitrogen.

Just as water can find its way downhill by many different pathways, so there are many chemical pathways by which SO_2 and NO_x are converted to acid in the atmosphere. The major pathways do not involve oxygen directly, but instead involve more powerful oxidizing agents, including hydroxyl radical, hydrogen peroxide, and ozone, which are ubiquitous in the atmosphere.

For both SO_2 and NO_x, a major reaction pathway is with hydroxyl radical (OH), a highly reactive molecule.[62] The reaction between SO_2 and (OH) produces the unstable bisulfite radical

$$SO_2 + (OH) \rightarrow (HSO_3)$$

This radical is immediately oxidized to acid sulfate. This reaction takes place in clear air ("gas-phase oxidation") producing an aerosol or haze of acid sulfate. The reaction goes faster in intense sunlight and is more important in summer and at midday. The pathway produces up to 20-25% of airborne acid sulfate. Acid haze is often partially neutralized by airborne ammonia to form ammonium acid sulfate, but there is so little liquid water in these aerosols that they usually remain extremely acidic — as low as pH 1.0 or less.[55] Acid haze returns to earth by several routes: Some of it is adsorbed directly by the ground; some of it is "updrafted" through clouds and dissolves in cloud droplets; and some of it is washed out of the air by falling raindrops. Under most conditions, the amount of acid produced by this pathway is proportional to the amount of SO_2 in the air; thus reducing the level of SO_2 in the air will proportionally reduce the amount of acid formed.[63]

NO$_2$ reacts with (OH) to form acid nitrate directly:

$$NO_2 + (OH) \rightarrow HNO_3$$

This reaction is considerably faster than the one with SO$_2$ and proceeds appreciably even at night. It is thought to account for most of the acid nitrate formed in the air. In polluted, urban air, NO$_x$ can react with organic matter to produce the important intermediate, peroxyacetyl nitrate (PAN), a pollutant which may be transported long distances before it is eventually converted to acid nitrate. In acid smog, half the nitrogen may be in the form of PAN.[62]

The majority of airborne acid sulfate appears to be formed in cloud droplets ("aqueous-phase oxidation"). SO$_2$ dissolves to form the bisulfite anion (HSO$_3$$^-$), which then reacts with hydrogen peroxide (H$_2$O$_2$) to form acid sulfate. The lower the pH the faster this reaction proceeds. At pHs above 5.0, reaction between HSO$_3$$^-$ and ozone (O$_3$) becomes appreciable and may become the dominant pathway for acid formation.[62]

The oxidation of SO$_2$ to acid sulfate is also catalyzed on the surface of fine particulates present in the plume from smokestacks. The reaction rate is relatively slow, however, and this pathway is now thought to account for a minor fraction of the acid sulfate formed in the atmosphere.

The relative importance of the above pathways will depend on many factors, including the ratio of NO$_x$ to hydrocarbons in the air, the availability of sunlight, the humidity, and the presence of other pollutants. Overall, the conversion of SO$_2$ to acid is complete within several hours to several days, while NO$_x$ is probably converted to acid within hours.[62,63]

Sources of sulfur emissions

More than 95% of the sulfur emitted in the eastern U.S. comes from man-made sources, while less than 5% comes from natural sources, which include decaying vegetation, volcanoes, and sea spray. The following is an inventory of the sources of sulfur oxides.

Man-made sources: The burning of coal and oil accounted for virtually all the sulfur oxides emitted from man-made sources in the eastern U.S., about 30.6 million tonnes (calculated as sulfate) in 1980. (Table 2).[64] (Emissions are estimated from the sulfur content of each type of fuel and the amount of fuel consumed.) Electric utilities contributed 71% of the total SO_2, with the majority of emissions coming from coal-fired power plants. The remainder came chiefly from industrial, commercial and residential combustion; transportation; smelters; and industrial processes. An additional 2.1 million tonnes entered the U.S. from Canada, principally emissions from metal smelters; and 1.2 million tonnes entered the region from the western U.S.[65]

Coal-fired power plants dominate the emission of SO_2 because much of the coal burned in the eastern U.S. is high in sulfur (greater than 3% by weight). Although there are thousands of individual sources of SO_2, just the top 20 coal-fired power plants emit 20-25% of all SO_2 in the eastern U.S.[37] (Table 3). These 20 plants account for an even larger fraction of acid sulfate in rain, because their sulfur is emitted from very tall smokestacks. (Emissions from tall stacks remain aloft longer and have more time to be oxidized to acid than do emissions from stacks of lesser height.) The 20 largest plants emit an amount of sulfur comparable to the sulfur that falls in all the acid precipitation on all states east of the Mississippi combined. (Compare Tables 2 and 3.) The largest coal-fired plants have been the focus of efforts to control acid rain.

In Canada, metal smelters are the largest emitters of SO_2, accounting for 45% of that nation's SO_2 emissions.[64] (During the smelting process, metal ores containing sulfur are roasted at high temperature, and the sulfur is driven off as SO_2.) The giant metal smelter at Sudbury, Ontario, owned by the International Nickel Company, is the largest single emitter of SO_2 in North America.

Electric utilities contribute only 16% of Canada's sulfur emissions, because Canada relies heavily on hydropower and nuclear power rather than on fossil fuel combustion. Canada emits more SO_2 per capita than does the U.S., but comparisons are misleading because of the very different sources of SO_2 in the two countries.

Biological sources: Anyone who has savored the acrid aroma of a salt marsh or swamp knows that nature is a potent emitter of sulfur compounds. A variety of sulfur-containing gases (such as hydrogen sulfide and dimethyl sulfide) are produced by the action of soil bacteria on rotting vegetation and on inorganic sulfate.[66] Once in the air, these sulfur compounds are rapidly oxidized to SO_2 and further oxidized to acid sulfate.

Biological sulfur emissions are measured by sampling the air at locations representative of different soil types. The total biological emission of sulfur in the eastern U.S. was estimated to be 357,000 tonnes per year (calculated as sulfate), based on field measurements at 37 representative sites throughout the eastern U.S.[67] That amount is equivalent to the emissions from one large coal-fired power plant. Coastal wetlands, comprising only 7% of the land area in the eastern U.S., contributed one-half the sulfur. The remaining sulfur came from inland soils, which constitute 93% of the land area.

Biological sources are an insignificant cause of acid rain for several reasons: They account for less than 5% of total sulfur emissions in the region. The sulfur is emitted at ground level and returns to earth relatively close to its source. Furthermore, biological sulfur emissions are part of the natural sulfur cycle and do not represent a net addition of sulfur to the environment: Sulfur that is oxidized to acid sulfate is reduced again by microbial action upon returning to earth — a process that *removes* acidity from the environment. Microorganisms of course cannot distinguish between naturally-derived and man-made acid sulfate, but the *amount* of man-made acid sulfate deposited exceeds the capacity of microorganisms in many soils to assimilate it.

Sea spray: Near the ocean, some of the sulfate in precipitation comes from sea spray, not pollution. This sulfur is already fully oxidized and contributes no acidity to precipitation. (In fact, sea water is slightly alkaline.) A technique for distinguishing between the sulfate derived from sea spray and the sulfate from man-made pollution was developed in the middle of the nineteenth century. It is based on the fact that the ratio of chloride to sulfate in seawater is constant. Since the major source of chloride in precipitation near the coast is sea spray, measuring the amount of chloride in precipitation allows a determination of the amount of sulfate in precipitation contributed by sea-spray. In this book, the values given for sulfate in precipitation do not include the "excess" sulfate contributed by sea-spray.

Volcanoes: Volcanoes are the very symbol of sulfurous fumes, but they contribute an insignificant amount of acid sulfate to precipitation in the eastern U.S., because they are remote and because their emissions are usually small compared to man-made sources. Mt. St. Helens emitted about 250,000 tonnes of sulfur (calculated as sulfate) during the year of its eruption, comparable to the emissions from one large coal-fired power plant.[68] During major volcanic eruptions, much of the sulfur is injected into the stratosphere, where it can travel several times around the globe before returning to earth.[69] The sulfur is diluted as it travels and contributes little to acid deposition at appreciable distances downwind. Increased fallout of acid sulfate following major volcanic eruptions has been detected in ice cores from Antarctica, but the fallout is slight and is measurable only because the natural background of sulfur deposition in Antarctica is very low.[59]

The fate of sulfur emissions

Can the sources of sulfur emissions described above account for all the sulfur observed to be deposited? That is, does "what goes up" equal "what comes down"? The answer is yes. The fate of sulfur emissions in the eastern U.S. is discussed below and summarized in Table 4. Virtually all of the sulfur returns to earth within several hours to several days. About one quarter of the sulfur emitted in the eastern U.S. returns to earth in precipitation within the region. Another quarter of the sulfur settles on the region when it isn't raining ("dry deposition"), in the form of unreacted SO_2 and acid sulfate particles. Roughly another quarter of the sulfur is carried by the wind to Canada. The remainder is carried out over the Atlantic Ocean, where it contributes to acid rain as far away as Bermuda.

Amount of sulfur in precipitation: Virtually all the sulfur deposited in precipitation is in the form of acid sulfate.[70] (Typically, less than 5% of the sulfur is dissolved SO_2, and this remnant is rapidly oxidized to acid after falling to earth.) The amount of acid sulfate deposited annually by precipitation in the eastern U.S. totaled 7.3 million tonnes in 1980 (Table 2).

The annual amount of sulfur deposited is relatively uniform from state to state (every state is within a factor of 2 of the average), in contrast to the wide variation in sulfur emissions among states. This uniformity is due to the large-scale transport and mixing of sulfur in the atmosphere.

Amount of "dry deposition": Both SO_2 and acid sulfate also return to earth when it isn't raining, a process called "dry deposition." Although public attention has focused on acid rain, dry deposition in some localities contributes as much acidity as acid rain.

Laboratory and field experiments have shown that SO_2 is adsorbed by materials on the ground (such as soil, leaves and building stone) and then is rapidly oxidized to acid sulfate.[71] Each material adsorbs SO_2 at a characteristic rate; for example, cement is a better adsorber that asphalt. Also, the rate of adsorption is proportional to the amount of SO_2 in the air.[71] The amount of sulfur adsorbed by a given plot of land will depend on many factors: the types of materials, the surface area of the materials (a "square meter" of lawn may have many square meters of surface in contact with the air), and the weather (wet surfaces may remove more SO_2 from the air than dry surfaces). Field measurements on small plots of land have yielded average values for the rate of dry deposition that appear to be in reasonable agreement with theoretical predictions.[72]

Using empirical parameters, it is estimated that 9.9 million tonnes of sulfur (calculated as sulfate) are dry deposited on the eastern U.S. annually.[65]

Amount of sulfur carried outside the eastern U.S.: The amount of sulfur carried by wind to Canada can be estimated knowing the average level of sulfur in the air, and the average northerly wind speed. It is determined to be 6.0 million tonnes of sulfur (calculated as sulfate) per year, about two to three times more than the amount of sulfur transported from Canada to the U.S.[65]

The amount of sulfur carried over the Atlantic is determined in the same way, using the average westerly component of the wind, and is estimated to be 11.7 million tonnes per year.[65] (An independent estimate can be made using rainfall data from over the Atlantic, and is consistent with the above estimate.)[73]

Total amount of sulfur deposited: The four pathways described above deposit a total of 34.9 million tonnes of sulfur (calculated as sulfate) per year.

The total amount of sulfur deposited agrees well with the total amount of sulfur emitted (Table 4). In fact, the agreement is better than one might expect considering the moderate uncertainties in some of the terms. This agreement indicates that we have not overlooked a major source of sulfur. *We conclude that man-made sulfur emissions are the chief cause of acid sulfate in rain in the eastern U.S. and that these emissions are also responsible for dry deposition of acid sulfate.*

Sources and fates of nitrogen oxides

More than 90% of the NO_x emitted in eastern North America comes from man-made sources. Fossil fuel combustion emitted

15.5 million tonnes of nitrogen (calculated as nitrate) per year in 1980.[64] Part of this was derived from the nitrogen in the fuel. For example, coal can contain more than 1% nitrogen by weight. In addition, any flame or high-temperature combustion allows nitrogen and oxygen in air to combine, producing additional NO_x. Approximately 56% of NO_x emissions came from "stationary sources," i.e., from smokestacks on power plants and industrial and residential boilers. The remaining 44% came from "mobile sources," including motor vehicles, planes, and trains.

Natural sources of NO_x include forest fires, lightning, microbial processes in soils and, to a lesser extent, oxidation of ammonia and input from the stratosphere. These natural sources accounted for a total of 1.8 million tonnes of nitrogen (calculated as nitrate) per year.[61]

Approximately 6.6 million tonnes is deposited annually in eastern North America as acid nitrate in rain and snow. Another 4.4 to 8.8 million tonnes is dry deposited. The remainder is carried outside the region.[61]

In summary, man-made emissions are by far the greatest source of SO_2 and NO_x in eastern North America, and are the chief cause of acid rain in the region.

4

THE TRANSPORT OF ACID POLLUTION

Few aspects of acid rain have provoked as much controversy as the question of which sources of pollution are responsible for acid rain in the Northeast and elsewhere in the eastern United States. As politically charged as the issue may be, determining "who is to blame" is the first step toward developing an efficient control strategy and resolving the complex issues of equity. Several general conclusions may be drawn: Each state causes some of its own acid rain through local emissions of SO_2 and NO_x. In addition, substantial amounts of pollution are transported hundreds, and even thousands, of miles downwind, across state and national borders. The task for scientists is to determine *how much* pollution is transported from one region to another.

Most states east of the Mississippi River emit as much sulfur as they receive in acid rain (Table 2). In principle, these states could cause their own acid rain. However, closer inspection of Table 2 shows that for some states, such as Vermont and Arkansas, the amount of sulfur deposited by acid rain far exceeds the sulfur emitted within the state. Acid rain in these states clearly cannot come from local sources but instead must be "imported" from polluters upwind. Furthermore, acid rain falls far out over the Atlantic — where there are no significant marine sources of sulfur — and gradually tapers off hundreds of miles at sea. Clearly, this pollution is transported from the East Coast. On Bermuda, six hundred miles offshore, storms coming from the East Coast bring highly acid rain, while storms coming from the open Atlantic to the east do not.[73]

Satellite photos reveal that acid haze can travel a thousand miles or more.[74] It has been argued that satellite photos do not rule out the possibility that a blob of haze may be continually renewed as it moves across the country, depositing and picking up pollution as it travels. However, acid sulfate haze is known from field and laboratory studies to be able to remain aloft for days, ample time to be transported hundreds of miles, and the haze remains intact even when it travels over areas with low emissions (see Plate II). The photos are strong direct evidence that air pollutants can be transported great distances.

Factors affecting transport

Air pollutants are carried downwind anywhere from a few miles to over a thousand miles, depending on several factors: the wind speed, the height of the smokestack, weather variables, and the chemical state of the pollutants.[75] One of the most important factors is whether the pollutants are emitted in the so-called "mixing layer" of the atmosphere — the layer of air closest to the ground in which there is good vertical mixing. The mixing layer is typically about 3,000 feet high and is often seen as the brownish blanket of polluted air covering a city. Pollutants emitted within the mixing layer make contact with the ground relatively quickly and are transported a distance that is roughly proportional to the height of the smokestack. But pollutants emitted above the mixing layer are effectively "decoupled" from the ground and can be transported great distances by strong winds aloft. This will happen when the smokestack is so tall that the hot plume of pollutants rises and "punches" through the mixing layer. It can also occur when the mixing layer is shallow, as often occurs at night or under certain atmospheric conditions.

The chemical state of the pollutants also affects the distance they are transported: Acid sulfate is transported further than SO_2 because it is less strongly adsorbed by the ground;[72] thus atmospheric conditions that favor the formation of acid will increase transport distance. Rain, of course, can wash pollutants to earth close to their source.

Quantifying the transport of pollutants

Three techniques have been used to determine how much sulfur pollution is transported from one region to another: computer simulation, back-trajectory analysis, and trace-element analysis. Each technique has confirmed the importance of long-distance transport.

Computer simulation: The distance that air pollutants are carried downwind can be estimated on a computer using data for the location of the emitters, the amount of pollution emitted and the direction and speed of the wind.[76] (More sophisticated computer programs also include the height of the smokestack, since wind speed varies with altitude.) The calculations also require estimates for how fast the pollution returns to earth. This will depend on how strongly SO_2 and acid sulfate interact with the ground, and will involve estimates for the relative abundance of SO_2 and sulfate in the air, which in turn depends on variables such as the amount of sunlight, the presence of other pollutants, the humidity, etc. (see Chapter 3). As a result of these uncertainties, computer simulations only provide estimates, not exact results. Nevertheless, the results from several computer simulations have agreed well with the experimental observations from back-trajectory analysis and other techniques (discussed below).

Computer simulations have shown that the majority of acid sulfate pollution in the Northeast comes from outside the region, chiefly from sources in the Ohio Valley/Midwest. For example, the "AIRSOX" model, developed at the Brookhaven National Laboratories, estimates that "87% of the sulfate in New York and New Jersey is due to (long-distance) transport and that 92% of the sulfate in New England is due to transport. The model further predicts that only 10% of New England's sulfate is caused by emissions in New York and New Jersey, with the balance of the sulfate resulting from emissions further upwind."[77]

According to the averaged results of seven computer models considered by the U.S./Canada Work Group on Transboundary Air Pollution, midwestern sources contribute more than 50% of the airborne sulfate pollution in the Adirondack Mountains, with approximately 20% of the sulfate coming from the Sudbury sector of Canada, and less than 10% coming from the Northeast itself[78] (Table 5). Other sites in the Northeast may be more heavily influenced by local sources of pollution. At Brookhaven, Long Island as much as 30% of sulfur deposition may come from upwind emissions in the New York metropolitan area.

Back-trajectory analysis: Whereas computer simulation starts with the sources of pollution and asks, "Where does the pollution wind up?", back-trajectory analysis asks the complementary question: "Where does the pollution at a given site come from?" In this technique, the chemistry of each rainstorm and snowstorm is monitored at a given site, and the trajectory of each storm is determined back in time for the previous 48 hours using weather data, satellite photos, etc. This information enables one to correlate the origin of the air mass bringing each storm with the amount of pollution it delivers. One can imagine a giant "pie" centered on the monitoring site, with different sectors of the pie delivering different amounts of pollution. It is found that sectors with the highest SO_2 and NO_x emissions deliver the most acidic rain.

An extensive study using this technique was carried out at Whiteface Mountain in New York's Adirondacks, sponsored by the electric utilties as part of the Multistate Atmospheric Pollution Power Production Study.[78] Over the course of a year, 62% of the sulfate in precipitation falling on the Adirondacks was deposited by storms coming from the Ohio Valley/Midwest sector, 25% came from the Sudbury sector of Canada, and less than 5% came from the Northeast itself (Table 6 and Figure 4). Similar results were observed for nitrate and for total acidity (Table 6). *These results indicate that in order to reduce acid rain in the Adirondacks, one must reduce emissions from the major sources of pollutants in the Midwest sector and Canada.* Even if all the SO_2 and NO_x originating in the Northeast were totally eliminated, that would have a neglible effect on acid rain in the Adirondacks.

Why does the Midwest deliver so much acid sulfate to the Adirondacks? The amount of acid sulfate delivered by a sector is the product of two factors: the amount of pollution in the rain, and the amount of rain. Midwestern precipitation is highly polluted: four times as concentrated in sulfate as precipitation from the Northeast (Figure 5). Furthermore, since the prevailing winds blow west to east, most of the precipitation in the Adirondacks comes from the Midwest — three times more than from the Northeast (Figure 6). Together, these two factors account for the 12-fold disparity (62% vs. 5%) between the Midwest's and the Northeast's contribution to acid sulfate in the Adirondacks.

Similar results have been obtained at other sites in the eastern U.S. and Canada. In central Massachusetts, highly acidic rain (pH 3.9 or less) was brought almost exclusively by weather from the Ohio Valley/Midwest, while less acidic rain (pH 4.8 or more) was brought by storms from over the Atlantic.[80] In southern Rhode Island, a six-month study revealed that storms coming from the midwestern states were highly acidic (pH 3.8 to 4.5) while storms from the Atlantic Ocean or the seaboard states were less acidic (pH 5.1).[81] In Haliburton-Muskoka, a popular vacation area in south central Ontario that is highly sensitive to acid rain, the majority of acid sulfate and nitrate was deposited by precipitation from the south and southwest octants, that is, from the Ohio Valley/Midwest.[82] Precipitation coming from the Sudbury sector (containing the metal smelters) was infrequent, but also contained very high concentrations of acid sulfate.

By itself, back-trajectory analysis cannot determine precisely where along the trajectory a storm has picked up pollution. However, the location of the major emitters is known from inventories developed by the Environmental Protection Agency and by others. Coupled with this additional information, trajectory analysis provides compelling evidence that acid-forming pollution is carried long distances from high-emissions regions such as the Ohio Valley/Midwest, to pristine regions hundreds of miles downwind.

Trace element analysis: Trace element analysis is an indirect technique that seeks to infer the sources of sulfur pollution by measuring other chemical elements emitted along with the sulfur. The technique is based on the fact that different pollution sources emit characteristic chemical "signatures" (characteristic amounts and types of chemical elements) that depend on the type of fuel burned. For example, the pollution from coal-fired plants is relatively high in selenium, an element chemically similar to sulfur; pollution from oil-fired plants is relatively low in selenium. By measuring the levels of such elements in polluted air, it is in principle possible to infer whether the pollution came predominantly from coal-fired sources (and therefore predominantly from the Midwest) or from oil-fired sources (and therefore chiefly from the Northeast). (A similar technique has been used to trace oil spills, based on the fact that the oil in each ship's hold is a chemically unique mixture).

Preliminary studies at Watertown, Massachusetts showed that high levels of acid sulfate pollution correlated with winds from the Ohio Valley/Midwest sector, and also contained high levels of selenium.[83] Conversely, winds from the east, i.e., from the city of Boston, contained relatively low levels of acid sulfate and selenium. This suggests that episodes of high acid sulfate at the Watertown site are due to air masses from the Ohio Valley (where sources are predominantly coal-fired) rather than to local sources (which are predominantly oil-fired). The technique was also used at sites in New York State and again implicated midwestern sources of pollution.[84]

On the basis of preliminary work using manganese and vanadium as tracers, Dr. Kenneth Rahn of the University of Rhode

Island reported that long-distance transport of pollutants from the Midwest to the Northeast was less important than previously thought.[85] However, recent work has shown that manganese and vanadium are not reliable tracers for determining the sources of pollution in the eastern U.S.: manganese levels may be higher in midwestern steel emissions and soil dust than in coal emissions.[86] Also, manganese is removed from the atmosphere faster than sulfur during long-range transport, leading to results that do not accurately reflect the transport of sulfur.

Transport of nitrogen oxides

NO_x emissions are thought to produce more localized effects than SO_2 since nearly half of all NO_x is emitted at ground level, by motor vehicles, and therefore would not be carried as far by the wind. Furthermore, NO_x emissions are converted relatively rapidly into acid nitrate, which is adsorbed by the ground more quickly than acid sulfate.

Yet even in rural areas precipitation contains high levels of acid nitrate, which often cannot be accounted for by local sources of pollution. In fact, the ratio of sulfate to nitrate in precipitation at rural sites in the eastern U.S. (e.g., at Hubbard Brook, New Hampshire) is about the same as the ratio of sulfur to nitrogen emissions in the eastern U.S. as a whole.[63] This indicates that nitrogen emissions are indeed transported long distances, as are sulfur emissions. One possibility is that the nitrogen is transported long distances as the intermediate, peroxyacetyl nitrate (PAN), before being converted to acid.[62] Another possibility is that the acid nitrate in precipitation in rural areas might originate preferentially from sources with tall smokestacks. Consistent with this is the observation that in the Adirondacks, acid nitrate levels are several times higher in precipitation coming from the Midwest than in precipitation from the Northeast, despite the high density of traffic in the Northeast. This suggests that NO_x emissions from tall stacks, rather than from cars and other mobile sources, might be the predominant source of acid nitrate in the Adirondacks.

Global-scale transport

Acid rain is becoming a global phenomenon. The acidity of snow in the Arctic, for example, has been increasing during the past 25 years.[87] "Arctic haze," consisting largely of acid sulfate, is thought to originate from large, coal-fired sources in central Russia, Europe and possibly North America.[88] (In the Northern latitudes, there is so little moisture in the cold air that precipitation is infrequent, allowing pollution to travel far north before being brought to earth.) Because Arctic haze modifies the amount of sunlight absorbed by the earth, there is concern that the haze might contribute to the "greenhouse effect" and thereby alter the earth's climate. There is also concern about the long-term effect of acid snow on the fragile Arctic tundra.

Acid rain on Bermuda originates with emissions from the eastern U.S.[73] Acid rain on Hawaii is believed to come in part from polluters in Asia.[89] The level of acid sulfate in the stratosphere due to man-made sources appears to be increasing by as much as 10% per year, and this may enhance transport of sulfur around the globe.[90]

5

THE EFFECTS OF ACID RAIN

"A lake is the landscape's most beautiful and expressive feature. It is earth's eye: looking into which the beholder measures the depth of his own nature."

Henry David Thoreau

Water is the most important substance on earth, and acidity is one of its key properties. Everything on earth — living and inanimate — is either composed of water or in contact with water at some time in its existence. It is not surprising then that acid rain has caused profound environmental changes, some benign, some alarming. The challenge to scientists has been to determine what changes have already taken place, and to predict what future effects will emerge. Acid rain is an intricate scientific puzzle that has taxed the entire range of physical and life sciences.

To date, the major adverse effects of acid rain include damage to lakes, streams, and forests; degradation of soils; leaching of toxic metals in the environment; damage to man-made materials; adverse respiratory effects in humans; and degradation of air quality. It is very difficult to predict the future effects of acid rain on something as complex as an entire ecosystem. We do not know whether most of the effects that will occur have already taken place, or whether we have only seen the tip of the "acid iceberg" to come.

Effects on lakes and streams

During the past several decades, acid rain in eastern North America has caused sensitive lakes and streams to acidify, leading to the decline and death of fish and other aquatic life. The extent of this damage in Scandinavia, Canada, and the U.S. is described in Chapter 1.

To appreciate the pressure that acid rain has put on sensitive lakes and streams, compare the acidity of rain with the acidity of lake water. Healthy lakes normally have a pH around 5.6 and above.[38] When a lake becomes as acidic as pH 5.0, adverse biological effects set in. At pH 4.5 or below, a lake is usually considered "dead" — incapable of supporting the rich variety of life found in healthy lakes. Yet precipitation in the eastern U.S. is several times more acidic still (average pH 4.3), and in the northeastern U.S. it is as acidic as pH 4.0. Thus if lakes and streams were simply huge puddles of rainwater, every lake east of the Mississippi River would be virtually devoid of life. What prevents this catastrophe is that acid rain is largely neutralized by the soils and rocks underlying most watersheds. The most common alkaline rocks are limestone and dolomite (calcium and magnesium carbonates), which are chemically similar to commerical stomach antacids.

However, large areas in the United States contain rocks that are poor buffers (such as granites, gneisses, and quartzites) and that give rise to poorly buffering soils. These areas include the northeastern U.S.; parts of the Appalachians, Smoky Mountains and southern states; the Boundary Waters Canoe Area in Minnesota, and parts of northern Wisconsin; and parts of the Pacific Northwest and the Rocky Mountains.

Figure 7 shows the regions in the U.S. where lakes and streams are especially sensitive to acid deposition: regions where surface waters have a total alkalinity of less than 200 microequivalents per liter. *Alkalinity* is a measure of the ability of a body of water to neutralize added acidity. It is defined in different ways by different investigators but is usually a measure of the concentration of bicarbonate or carbonate in the water.

An alkalinity of 200 microequivalents per liter is arbitrarily taken as a cutoff for sensitivity to acid rain[91]: If a body of water with this alkalinity is mixed with a roughly equal volume of acid precipitation — typically containing about 100 microequivalents of acid per liter (pH 4.0) — then the total alkalinity would be reduced to about 100 microequivalents per liter. This is the brink of acidification — the level below which the body of water acidifies rapidly and biological effects are observed. The cutoff point is somewhat crude because it does not take into account the amount of acid precipitation that falls, the ability of the surrounding watershed to neutralize acid precipitation, or how fast the body of water regenerates alkalinity.

Lakes at high altitudes are particularly sensitive to acid rain. Normally, only 5 or 10% of the water in a lake comes from rain that has fallen directly on the lake; the great majority of water is from rain that has fallen on the surrounding watershed and run off into the lake. Thus most of the water in a lake has had an opportunity to be neutralized by soil. But lakes at higher altitudes are often nestled in rocky outcroppings, and comprise a large fraction of their whole watershed. In extreme cases, the rainwater runs untempered over a few bare rocks, directly into the lake. The problem is even worse in winter and spring when a carpet of snow or ice prevents rain and meltwater from contacting the ground at all before running into a lake. In that case the lake receives a large pulse of acid which can kill fish and other aquatic life outright.

Are lakes getting more acidic?

A few researchers have questioned whether the acidity of lakes is in fact increasing.[92] Historical records must be interpreted with caution because a number of errors can be introduced when sampling, storing and analyzing lake water that can cause a lake to *appear* to have been less acidic in the past than it really was: The acidity and chemistry of a lake can change significantly during the course of a day as aquatic organisms photosynthesize and respire, giving off and taking up carbon dioxide. Thus water samples taken at one time of day or season may not reflect the lake's *average* acidity. Water samples may have been stored in containers that affected their acidity. Outdated analytical techniques may have overestimated the acidity or underestimated the alkalinity of a lake.

Despite these cautions, there is now a large body of convincing evidence that several thousand lakes and streams have become more acidic in Ontario, Nova Scotia, New Hampshire, Vermont, Maine, New York, New Jersey, Pennsylvania, North Carolina, and Florida (for a critical review, see Ref. 91). In the Adirondacks 25% of the nearly 1,000 lakes sampled have acidified to the extent that they no longer support game fish. In the western U.S., acidification has been reported in the Sierras[93] and in the Colorado Rockies.[34]

Some researchers have also questioned whether the trend toward acidified lakes is the result of acid deposition or other causes.[94] For example, naturally acidic lakes can be found in the eastern U.S. These are generally clear lakes with a telltale brown color. (The color is due to organic acids, such as the tannins which give tea its distinctive hue.) They include peatland lakes in the Northeast and cypress swamps in the South. However, naturally acidic lakes are distinct from the so-called "oligotrophic" lakes which are the focus of concern in the U.S., Canada, and Scandinavia.[95]

Factors other than acid deposition can cause a watershed to become more acidic. For example, fertilizers produce acidity when the ammonium they contain is oxidized to acid nitrate. Thus runoff from farms can cause adjacent lakes and streams to acidify. The growth of trees on previously agricultural land can greatly increase soil acidity. (The reason is that when trees and other vegetation take up calcium, magnesium and other positively charged ions from the soil, they maintain the electrical neutrality of soil by secreting hydrogen ions, i.e., acidity.) Lowering the water table can expose both organic matter and minerals that contain sulfur to the air. As these are oxidized, they produce organic and sulfuric acids. However, these "land-use changes" do not appear to explain the large-scale acidification observed in the U.S., Canada, or Scandinavia.[7] In the high-altitude lakes in the Northeast, and in areas in Ontario that have never been logged, it is clear that lakes have acidified due to air pollution and not to other forms of human intervention. In the studies cited above and reviewed by scientists for the U.S. Environmental Protection Agency, acid deposition was concluded to be the dominant cause of acidification.

As discussed in Chapter 1, fish have been lost from several hundred acidified lakes and streams in the Adirondacks, Ontario and Nova Scotia, as well as in Scandinavia.[96] The death of fish illustrates a particularly insidious aspect of acid rain: the effects of acidity can be synergistic. For example, acidity can kill fish by interfering with the fish's salt balance;[97] by causing reproductive abnormalities[7]; by leaching aluminum into the lake at levels toxic to fish gills, so that the fish literally suffocate[98]; and by killing the organisms on which fish feed. The combined effect of these stresses can wipe out a fish population.

The loss of fish also puts pressure on other animals higher up the food chain: osprey, otters, and other birds and mammals.[13] It has been suggested that the toll of acid deposition on aquatic systems is not best measured by the number or particular kind of species being lost, but by the fact that whole ecosystems are being altered or destroyed, in some cases irrevocably.

Effects on forests

A new sense of urgency has been kindled by recent evidence that acid rain may be responsible for the widespread decline in the growth of evergreen forests in the eastern U.S. and Western Europe during the past several decades. Severe damage to forests in West Germany has already prompted that country to undertake a major reduction in SO_2 emissions.

Laboratory and field studies have shown that acid deposition can 1) damage leaves, roots, and microorganisms that form bene-

ficial symbiotic associations with roots; 2) impair reproduction
and the survival of seedlings; 3) leach nutrients such as calcium
and magnesium from soils; 4) dissolve metals in the soil such as
aluminum at levels potentially toxic to plants; and 5) decrease a
plant's resistance to other forms of stress, including pollution,
climate, insects, and pathogens.[99]

The field remains speculative and controversial in part
because it is virtually impossible to recreate the natural environment of forests in a laboratory. This is particularly true of the
most seriously affected forests in the U.S.: high mountain forests
which are already highly stressed from wind and cold. Furthermore, trees have too long a life cycle to study adequately in the
laboratory. Subtle changes that might require a human generation or more would go unnoticed in brief laboratory experiments.
Thus relatively little is known about the long-term physiological
responses of trees to acid rain.

It is now certain that there is a widespread and unprecedented
decline in the growth of high-altitude forests in the eastern U.S.,
primarily affecting evergreens but also affecting some hardwoods. This decline extends from northern Vermont to as far
south as Tennessee. According to Arthur Johnson, at the University of Pennsylvania, the "very rapid shift to abnormally small
(tree) rings is widespread, substantial and sustained, and would
appear to have important implications for ecosystem stability."[22]

By far the greatest damage has been observed at Camels Hump
in Vermont's Green Mountains, where 50% of the red spruce
have died and the number of spruce seedlings declined by at least
50% since 1965.[18] (See Plate I.) A recent survey failed to find a
single healthy seedling.[100] Maples and beeches, while less
dramatically affected, have also declined. The spruce forest at
Camels Hump spends about one third of its existence shrouded
in fog and mist which is often highly acidic.

The physiological cause of death of spruce at Camels Hump is
not known. Analysis of tree cores has shown that the level of

aluminum — an element known to be potentially toxic to plants — has tripled since the 1960s. Other potentially toxic metals such as vanadium have increased in concentration as well. These metals become more soluble with increasing acidity and presumably have been leached from the soil during the past few decades by acid deposition. However, it is not known whether the metals are directly responsible for the death of the spruce. An imbalance in the ratio of calcium to aluminum in the soil has been hypothesized to damage fine root hairs, though the evidence is equivocal.[22] The root system of the spruce was also found to lack the nodules containing symbiotic fungi — called mycorrhizae — which are normally present and which are thought to aid in the uptake of water and nutrients. Insect pests, climate changes and other natural factors have been ruled out as plausible causes. The damage has affected trees and forest stands of all ages. Hubert Vogelmann of the University of Vermont has hypothesized that acid deposition increases the trees' susceptibility to drought and other stresses, acting either directly on the mycorrhizae or indirectly through increased aluminum levels. It has recently been suggested that the spruce may be "overfertilized" by the high levels of nitrate in acid precipitation.

The severe damage at Camels Hump illustrates that an initially small change in an ecosystem can be compounded into catastrophic results.[100] As mature spruce trees die and fall, the forest floor becomes increasingly exposed to fierce winter winds which uproot shallow-rooted spruce seedlings. The winds and sunlight dry out the forest floor, increasing the stress on seedlings. Sunlight on the forest floor spurs the growth of ferns that compete with the few remaining spruce seedlings. As mature trees fail to be replaced by new trees, there are fewer seeds available to propagate the species. As a result, there are now virtually no viable spruce seedlings on Camels Hump. A venerable species has come to a halt on the mountain, a species whose stately members used to live an average of 150-200 years and in some case upwards of 300 years. The damage is irreversible for generations.

Decline in the growth of evergreens during the past 20-30 years has also been documented in the Adirondacks,[20] in the New Jersey Pine Barrens,[21] and in Maine, based on an analysis of tree rings. In the Adirondacks, balsam fir as well as red spruce have been affected. In the Pine Barrens, the decline and death of loblolly, shortleaf and pitch pine correlates with the increasing acidity of streamwater in the area.[101]

The decline in growth at these widely separated sites was initially thought to have been triggered by a severe drought in the eastern U.S. in the mid-1950s. More recent work, however, indicates that the decline began earlier and affected areas known not to have suffered a drought. Also, the growth of trees would have been expected to resume after the drought. As other natural factors have been eliminated, acid deposition remains as the suspect cause. Recent observations of red spruce decline on Mt. Mitchell in North Carolina and elsewhere in the Great Smoky Mountains tend to confirm this.

Pine trees in the Great Smoky Mountains have virtually stopped growing during the past 20-25 years according to an analysis of annual growth rings by researchers from Oak Ridge National Laboratory.[102] During the same two decades, regional SO_2 emissions from fossil fuel combustion increased by about 200 percent. The connection between SO_2 emissions and growth suppression is made compelling by the finding that a similar decline in growth occurred between 1863 and 1912, "a period of smelting activity and large sulfur dioxide releases at Copperhill, Tennessee, 88 kilometers upwind." According to the researchers, the smelter "contributed to the destruction of all vegetation with 16 km of Copperhill." The study found that the recent decline in growth was paralleled by increased concentrations in trees of aluminum, cadmium, copper, zinc, and other metals potentially toxic to plants. These metals are dissolved from soils by acid deposition, and they are also emitted in significant amounts by fossil fuel combustion itself.

Trees in the Ohio Valley/Midwest have also been damaged by air pollution, according to Dr. Orie Loucks, director of the Holcombe Research Institute at Butler University in Indianapolis.[103] Loucks cited "discolored foliage on white pine, poplar, sycamore, maples and other trees in cities"; an "unusually severe narrowing of tree rings in these areas that began in the mid-1960s"; and "an unusual degree of mortality in the trees in cities of the Ohio Valley." Though Loucks hypothesized that airborne acid sulfate in combination with ozone pollution were major factors in the damage, other causes have not been ruled out.

The forest industry has not voiced strong concern about the potential threat of acid rain to commercial timberland. In part, this is because commercial stands of forest do not appear to have been affected by acid rain; in part because commercial forests are already heavily managed and are therefore considered less sensitive to acid deposition; and in part because the forest products industry itself is a significant emitter of air pollution. Nevertheless, even a small decline in forest productivity would have a significant economic effect.[104]

The situation is even more alarming in Central Europe, where several million acres of forests have been damaged by air pollution. In Germany, it was estimated in 1982 that over 7% of trees in all forests were affected, including evergreens and hardwood species.[105] By 1983, that estimate had climbed to 34%, in part reflecting increased damage and in part reflecting better reporting of damage.[104] A wide variety of symptoms have been reported, including some (such as the active dropping of green leaves and the formation of calcium sulfate crystals within leaves) that have rarely or never before been described. Four working hypotheses have been advanced.[106] The damage may be caused by: 1) the direct action of gaseous pollutants, particularly SO_2 and ozone, which are known to be quite toxic to vegetation; 2) severe magnesium deficiency brought about by the leaching of magnesium from leaves and soils; 3) "general stress", including a decline in photosynthesis and the accumulation of toxic metabolites, brought about by air pollutants; and 4) increased concentrations of aluminum in the soil mobilized by acid deposition. (For example, at one site in West Germany that is downwind from heavy industrial emissions in the Ruhr Valley and that receives large amounts of wet and dry acid deposition, the extensive death of evergreens and beech trees during the past decade has been paralleled by a tenfold increase in the level of aluminum in the soil.[107])

There are important differences between the forest damage in Europe and that in the United States. In the U.S., damage has so far been restricted to trees growing at high altitude (which spend much of their time in highly acid cloud water), and the chief symptom is "dieback", or death of leaves, starting from the top of the tree. In Europe, trees at all altitudes have been affected, and the damage appears to affect the entire tree. Whatever the differences, it should be emphasized that all of the hypotheses advanced to explain the damage to forests in both Europe and the United States implicate SO_2 or NO_x emissions and the secondary pollutants they produce, namely acid sulfate, acid nitrate, and ozone.

An eight-year study of acid deposition in Norway found slightly *increased* forest growth in regions deficient in nitrogen and sulfur, despite relatively high levels of soil aluminum.[7] Thus nitrogen and sulfur in acid deposition may help fertilize forests in areas normally deficient in these nutrients. However, the study found slightly *decreased* forest growth in regions where nitrogen and sulfur were not limiting nutrients. Acid deposition is likely to be detrimental to these Scandinavian forests in the long-term due to the leaching from the soil of other nutrients, such as calcium and magnesium (see below).

Acid rain also adversely affects vegetation on the forest floor. In laboratory experiments, simulated acid rain was found to inhibit fertilization in ferns,[108] and to inhibit nitrogen fixation in algae growing on the surface of forest soils.[109] Acid rain renders some plants more susceptible to pathogens such as fungi, perhaps by damaging the cuticle or other protective surface of the plant. Other plants become less susceptible, presumably because acid rain renders the pathogens themselves less virulent. The extent and significance of these effects in the environment is not yet known. A report from the Office of Science and Technology Policy warned that long-term changes in soil microorganisms, especially those involved in decomposing litter and recycling nutrients, may eventually be the most serious threat of acid deposition. "The evidence that increased acidity is perturbing populations of microorganisms is scanty," the O.S.T.P. report concluded, "but the prospect of such an occurrence is grave."

Effects on soils

It has been argued that forests in the Northeast and Canada are growing on naturally acidic soils for which acid rain should not make any difference.[94] On the contrary, acid rain can adversely affect soils that are already naturally acidic. To understand why, it is necessary to consider the behavior of ions in soils. Most of the ions of interest are positively charged: these include acidity (hydrogen ions), nutrients (such as calcium and magnesium ions), and potentially toxic metals (such as aluminum, lead, mercury, cadmium and other metal ions). These positive ions are normally not free to migrate through the soil because they are tightly bound to the negatively charged surface of large, immobile soil particles. (The negative charges are provided by silicates on the surface of clays, and by organic acids on particles of organic matter.) The ability of a soil to bind positively charged ions is called its "cation-exchange capacity" and is responsible in large part for mitigating acid deposition. The continued addition of acidity can deplete the cation-exchange capacity of a soil.[110]

Acid rain "mobilizes" the positive ions in soils in two ways. First, the hydrogen ions in acid rain displace the other positive ions from their binding sites, increasing the concentration of these ions in the soil water. When the soil particles are no longer able to bind any more hydrogen ions, the concentration of hydrogen ions in the soil water will also increase. Second, the sulfate and nitrate ions in acid rain, being negatively charged, act as mobile "counterions" which allow the positive ions to be leached from the soil. In short, acid rain unlocks the acidity, nutrients and toxic metals bound to soil particles.

Although there are variations on this scenario,[111] the cumulative degradation of soils appears to be a real phenomenon. Even in lakes that have not yet begun to acidify, the concentration of calcium and magnesium in the water has been found to increase with time, indicating that these nutrients are indeed being leached from surrounding soils.[95] The level of sulfate in acidified lakes is often high. Furthermore, the concentration of aluminum, vanadium, and other potentially toxic metals has increased in both soils and lakes over the past few decades, and these elements have been incorporated by plants and animals in increasing quantities.

Effects on crops

Most cropland receives far more acidity from fertilizer than from acid deposition. The potential acidifying effect of fertilizer applied at 70 kg of nitrogen per hectare is about 10 kiloequivalents acidity/hectare — an order of magnitude more acidity than contributed by acid deposition.[110] In laboratory experiments, artificial acid rain has been shown to damage the leaves of a variety of crops, but usually at a pH below 3.0, well below values typical of current rainfall. The direct effects of gaseous pollutants, such as ozone and SO_2, again appear to be of more concern than the effects of acid rain. Ozone is estimated to damage several billion dollars worth of crops annually, mostly soybeans. The sulfur and nitrogen in acid rain help to fertilize some crops grown in nutrient-poor soils, but it is doubtful that highly polluted rain is essential for productive crop yields.

Health effects of toxic metals

Subtle threats to health sometimes turn out to be the most dangerous, because they can go undetected for a long time and affect large numbers of people. A growing concern is that acid rain is dissolving mercury, lead, cadmium and other toxic metals in the environment, and carrying them into drinking water supplies and into the food chain. These metals can damage the nervous system at relatively low concentrations, especially in children. To date there have been no reports of acute toxic metal poisoning attributable to acid deposition. However, the clinical effects of toxic metals are insidious and cumulative. Toxic metal poisoning therefore may pose a substantial, chronic public health problem even in the absence of acute manifestations.

Acidic drinking water leaches lead and copper from pipes as well as from the lead solder at the joints of non-lead pipes. In a study of drinking water drawn from acidified wells in the Adirondacks, public health officials in New York State reported at least two cases of elevated lead levels in the blood of children whose water supplies had pH's from 4.2 to 5.0 and which were served through lead pipes.[24] Even when the pipes were well flushed, the water contained 0.2 mg lead/liter — 4 times the federal standard for drinking water. Elevated copper levels were found in water drawn from a well with pH 4.95 and served through copper pipes. According to G.W. Fuhs, director of New York State's Environmental Health Center, "natural spring water in certain locations can contain copper up to 1 mg/liter and lead up to 0.2 mg/liter. In other words, concentrations of copper and lead equal to the U.S. drinking water standards can be leached by acid rainwater from natural rock and soil formations alone." In all, at least six counties in New York State have been found to have drinking water with lead levels that exceed the federal standard.

A comprehensive survey of drinking water in New York and New England found that when water was allowed to stand overnight in household plumbing, lead levels exceeded the E.P.A.'s Maximum Contaminant Levels in 8% of households surveyed.[23] More than 40% of the water samples exceeded the Secondary Maximum Contaminant Levels for copper. In one county in western Pennsylvania receiving highly acid rain, 16% of the households sampled whose tapwater came from cisterns had lead levels exceeding the U.S. standards for drinking water. In more than 40% of the surface waters in New York and New England, the concentration of aluminum is ten times higher than the limit for aluminum in kidney dialysis water — a finding of concern to dialysis patients.[23] (Aluminum can cause a syndrome known as "dialysis dementia", characterized by speech disorder, dementia, convulsions, and death.) High levels of aluminum have been found in the brains of patients with Alzheimer's disease, a degenerative disease of the brain, and recent studies suggest that aluminum may be a causal factor.[112]

Many toxic metals are spewed into the air during fossil fuel combustion. For example, mercury, cadmium, and other toxic metals are emitted during the burning of coal. Lead is emitted principally from motor vehicles using leaded gasoline.[38] These pollutants are now accumulating in the environment. Analysis of sediment cores from the bottom of acidified lakes in the Adirondacks has shown that the level of trace metals (and other pollutants such as polycyclic aromatic hydrocarbons, which cause cancer) has increased three- to four-fold over the past thirty years, with lead showing the most pronounced increase.[113]

Toxic metals dissolved by acid rain have found their way into the food chain. Fish containing levels of mercury exceeding federal standards have been found in acidified lakes in New York, Maine, Canada, and Scandinavia.[7,25] An added concern is that the species of fish with the highest mercury levels — trout and pike — are the fish that tend to be eaten in large quantities.[114] Recent studies of Quebec's Cree Indians, who eat large quantities of fish from acidified lakes, found elevated body burdens of mercury along with "very mild" symptoms of mercury poisoning.[115]

The accumulation of mercury in fish again illustrates that even a small increase in acidity can be compounded to produce serious consequences: As a lake becomes more acidic, mercury becomes more soluble and the concentration of mercury in the water rises; at higher acidity, mercury is preferentially converted by microorganisms to its more toxic form, methyl mercury; finally, acidity appears to increase the rate of uptake of methyl mercury by fish.[115] As a result, a relatively small increase in acidity can lead to a proportionately much higher level of mercury in fish.

Respiratory effects

For well over a century, acid haze, mist and fog have been suspected of exacting a toll on human health. In 1872, the British chemist R.A. Smith attributed the "great mortality" in Glasgow in part to the "remarkably high" level of airborne acid sulfate and other air pollutants.[5] In the late 1940s and early 1950s, "killer fogs" in Pennsylvania and London confirmed that severe air pollution can be lethal. Yet it has proven remarkably difficult to quantify the long-term health effects of low levels of acid haze.[117]

A widely publicized report has blamed air pollution for as many as 50,000 premature deaths annually in the U.S.[118] This figure is based on two types of epidemiologic studies: those comparing the mortality in two cities having different levels of air pollution, and those comparing the mortality within a given city on different days having different levels of air pollution. Such studies reveal a slight but unavoidable correlation between air pollution and premature deaths; more people tend to die on polluted days or shortly thereafter. However, several cautions are in order. The lower limit of the "confidence interval" is small, i.e., it is possible that there is a negligible effect of air pollution. Also, premature deaths are almost certainly among those who are already on death's doorstep — the very old and very ill. Furthermore, such studies do not distinguish between acid sulfate and the other components of air pollution, such as nitrogen oxides and ozone, which are also respiratory irritants. Animal studies have demonstrated that acid aerosols can affect lung function — for example, by increasing the time required for the lung to clear foreign matter. However, most laboratory studies on humans employ healthy, college-age volunteers, whereas those who would be most at risk to air pollution, the elderly and infirm, cannot be subjected to acid aerosols for obvious ethical reasons. It is therefore very difficult to get a good measure of the exact toll of acidic air pollution.

Effects on materials

"Pretty soon you won't be able to read the name on your grandma's tombstone," lamented one congressman, referring to the slow dissolution of marble, limestone and other materials by acid rain. Acid rain, haze and other acid-forming pollutants accelerate the corrosion of metals and paints, dissolve building stone, and degrade materials such as textiles, ceramics, leather, and rubber. SO_2 and its acidic products have caused well-publicized damage to such monuments and art treasures as the Taj Mahal, the Acropolis, Cologne Cathedral, and even the U.S. Capitol Building.[119] In Europe, the cost of corrosion attributable to SO_2 emissions was estimated by the Organization for Economic Cooperation and Development to be several billion dollars per year.[120]

Skeptics point out that the corrosion of steel and other materials is often caused by the direct effects of SO_2 itself, not by acid rain. It has proven difficult to distinguish the damage caused by acid rain from that caused by SO_2, but the distinction is important: SO_2 is already regulated by the Clean Air Act in part on the basis of so-called "welfare effects," including damage to materials. It has been argued that stricter control of SO_2 is not necessary. However, a sojourn to almost any town square in the East will reveal limestone monuments or bronze statuary that clearly have been degraded by acid rain rather than gaseous pollutants, the damage being more severe in those parts of the surface in contact with rain. Furthermore, acid rain has extended the corrosive effects of sulfur pollution beyond cities to remote and pristine areas.

Decreased visibility

"I never trust any air I can't see," quipped New York's former mayor, John Lindsay. During the past three decades, acid haze has steadily reduced visibility throughout the eastern U.S., from Louisiana to Maine. If there were no man-made pollution, visibility in the eastern U.S. would be expected to be about 25 to 40 miles. Now it has declined to as little as one mile during high pollution "episodes." (By comparison, visibility in Wyoming and other western states often exceeds 50 miles.) Acid haze is produced when SO_2 is oxidized in clear air to form microscopic particles of acid sulfate, which scatter light and reduce visibility (see Chapter 3). The haze is a complex brew consisting mostly of acid sulfate (often partially neutralized by ammonia), various organic compounds, and soot (elemental carbon). In the Great Smoky Mountains, haze was found to contain as much as 95% sulfate by weight.[121]

Haze is most pronounced in summer, when intense sunlight speeds up the photochemical reactions that produce acid sulfate; however, the decline in visibility has affected all seasons. In the Northeast, haze is at is worst when air has stagnated over the midwestern states, picked up SO_2, and then moved slowly northeastward. Time-lapse photos taken in the Green Mountains of Vermont dramatically show what happens when pollution-laden air moves into the region after stagnating in the Midwest[122]

(Plate II). The state of Maine recently joined a suit alleging that midwestern SO_2 emissions contribute to the "significant deterioration" of air quality in Acadia National Park (see Chapter 8). Acid haze is not confined to the eastern U.S.: even the Grand Canyon has been threatened by acid haze originating with SO_2 emissions from southwestern metal smelters.

STRATEGIES FOR REDUCING ACID RAIN

Adding lime to lakes and streams

Lakes and streams can be temporarily protected from acid rain by adding alkaline materials such as limestone (calcium carbonate) or lime (calcium oxide). Liming has been successful on a small scale over the past twenty years on Cape Cod, where fish have been maintained in otherwise acid lakes and ponds;[123] it has also been used experimentally in the Adirondacks,[124] and in Sweden,[125] which is spending 13 million dollars annually on its liming program.

Unfortunately, liming poses a number of problems that limit its effectiveness. It is difficult to lime streams, since large pieces of lime dissolve too slowly, while small pieces are swept into isolated pools and crannies. Liming a lake can cause a sudden surge of alkalinity, which can harm aquatic organisms. Some commercial grades of lime contain impurities, including toxic heavy metals. Also, many of the most sensitive, high altitude or remote lakes are simply inaccessible for routine treatment. Liming forests and entire watersheds is impractical. In short, liming is a stopgap measure that can address only a fraction of the acid deposition problem.

Reducing acid rain

The long-term solution to acid rain is to reduce the SO_2 and NO_x emissions that are the source of the problem.

Several science advisory panels have recommended reducing

the acidity of rain by roughly 50% in sensitive regions. For example, a 1981 report from the National Academy of Science concluded, "It is desirable to have precipitation with pH values no lower than 4.6 to 4.7 throughout [sensitive] areas, the value at which rates of degradation are detectable by current survey methods. In the most seriously affected areas (average precipitation pH of 4.1 to 4.2), this would mean a reduction of 50% in deposited hydrogen ions."[38] (An increase of 0.5 pH unit corresponds to a 70% decrease in acidity, not 50%; however the latter value has been widely accepted as the Committee's recommendation.) In some areas of the Northeast, the average acidity of precipitation is below pH. 4.1, so that an even greater percentage reduction would be required.

The Canadian members of the U.S./Canada Work Group on Transboundary Air Pollution recommended that "deposition of sulphate be reduced to less than 20 kilograms per hectare per year in order to protect all but the most sensitive aquatic ecosystems in Canada. In those areas where there is a high potential to reduce acidity and surface alkalinity is generally greater than 200 microequivalents per liter, the Canadian members recognize that a higher loading rate is acceptable."[26] This corresponds to reducing acid sulfate deposition by 30-70%.

The Swedish Ministry of Agriculture concluded that sulfate deposition of 1.5 grams acid sulfate per square meter per year "could be tolerated without entailing any risk of large-scale acidification damage. If we wish to prevent the acidification of even the most susceptible lakes and watercourses, the sulfate deposition will have to be reduced to not more than 0.9 g sulfate/sq. meter/year."[125]

By how much should emissions be reduced?

A few scientists have suggested that reducing SO_2 emissions by 50% "might not" reduce airborne acid sulfate by 50% — in fact, "might not" reduce acid sulfate at all. However, a variety of laboratory and field evidence indicates that reducing SO_2 emissions will indeed reduce airborne acid sulfate roughly commensurately: First, the amount of acid sulfate produced via the major chemical pathways is found to be *proportional* to the amount of SO_2 in the air, provided that there are sufficient levels of chemical oxidants in the air;[26] preliminary measurements of oxidant levels suggest that this is the case.[63]

Second, regional trends in SO_2 emissions are consistent with trends in acid sulfate levels. For example, a sharp increase in acid sulfate deposition in the South, particularly Florida, paralleled the sharp increase in SO_2 emissions in that region.[33] A decrease in SO_2 emissions in the Midwest paralleled a decrease in acid sulfate deposition at a downwind monitoring site at Hubbard Brook, New Hampshire during the same period.[127]

Third, the highest concentrations of acid sulfate in rain are observed to come from regions with the highest SO_2 emissions. For example, in the Adirondacks, the concentration of acid sulfate in precipitation was found to be several times higher in storms coming from regions with high SO_2 emissions (such as the Midwest) than in storms coming from lower emissions regions (such as the Great Lakes or New England). (See Chapter 4 and Figure 5.) This indicates that, on average, airborne acid levels are closely tied to emissions; reducing emissions is expected to reduce acid levels.

Fourth, during a copper smelter strike in Arizona and New Mexico in 1980, SO_2 emissions fell by approximately 85 to 90%. As a result, airborne acid sulfate levels dropped 50 to 90% below the levels found the previous summer.[128]

Fifth, "mass balance" arguments can place lower bounds on the proportionality between emissions and acid formation: Approximately 40-50% of SO_2 emissions are converted to acid sulfate before reaching the ground (including acid exported from the eastern U.S.).[65] Therefore, reducing SO_2 emissions from a typical source by, say, 90% must reduce airborne acid sulfate by at least 75%.

Finally, the ratio of SO_2 emissions to NO_x emissions is similar to the sulfate/nitrate ratio in precipitation. Since NO_x is known to be rapidly oxidized to acid, this suggests that SO_2 must also be quickly oxidized, i.e., that the formation of acid is not limited by the level of atmospheric oxidants. A 1982 report from the National Academy of Sciences concluded that "there is no evidence for a strong nonlinearity in the relationships between long-term average emissions and deposition."[63]

In summary, a decrease in SO_2 emissions will produce a roughly commensurate decrease in acid sulfate levels downwind.

Strategies for reducing emissions

Near-term methods for reducing SO_2 and NO_x emissions include fuel switching, coal cleaning, and emissions controls:

Fuel switching: Switching from high- (greater than 3% sulfur) to low-sulfur coal (less than 1% sulfur) can reduce emissions significantly, and is expected to account for approximately one third of the emissions reductions in a full-scale control program. The economic consequences of fuel switching are discussed in Chapter 1.

Coal cleaning: About half the sulfur in coal is in the form of iron pyrite, a dense mineral that can be physically removed by crushing and centrifuging the coal.[129] (The rest of the sulfur is chemically bound to carbon in the coal and is much more expensive to remove.) The process is about 50% efficient and thus can reduce SO_2 emissions by no more than 25%. Because of this limitation, and because many high-sulfur coals in use are already cleaned, the U.S. Department of Energy estimates that "coal cleaning cannot be used to control more than approximately 2 million tonnes of sulfur dioxide" — a 10% reduction.[30]

A variety of technologies are available for removing SO_2 and NO_x during or after combustion.

Scrubbers: The chief method for removing sulfur is flue-gas desulfurization — "scrubbing" — in which SO_2 in the stack gases reacts with a slurry of limestone and is removed as calcium sulfate. The process removes up to 90% of the sulfur. There are currently about 88 scrubbers in use, servicing 12% of electric generating capacity in the United States, and 40 more are under construction.[130]

Limestone-injected multistage burner: Particularly promising are staged combustion technologies that reduce both SO_2 and NO_x emissions. In the Limestone-Injected Multistage Burner (LIMB), fuel is burned in a series of steps.[131] The lowered combustion temperature reduces NO_x formation, while SO_2 is removed by reaction with dry limestone. The LIMB process can reduce both SO_2 and NO_x emissions by at least 50%, and it is several times more cost-effective than scrubbing. Another advantage is that LIMB requires less than one quarter the space of a scrubber and can be more easily "retrofitted" onto existing plants.

Fluidized bed combustion: In fluidized bed combustion, crushed coal is fed into a bed of dry limestone and the mixture suspended by jets of air to form a fluid.[132] Sulfur is removed as calcium sulfate. The high heat-transfer efficiency of the method allows lower operating temperatures, and thus reduced NO_x formation. Another advantage is that the process works with a wide variety of fuels, including refuse. Two fluidized bed combustion plants are currently operating in West Germany.

Long-term strategies: Virtually all the sulfur in coal can be removed by chemical treatment during *coal liquefaction and gasification,* but these processes are not yet economically attractive on a large scale. *Energy conservation* can reduce the demand for electricity and therefore for fossil fuel combustion. Increasing use of nuclear, solar, and wind power can offset the demand for fossil fuels, but these *alternative energy sources* are not expected to make major inroads in the eastern United States in the next several decades.

PART 3

POLITICS AND ACID RAIN

7

ACID RAIN AND THE POLITICS OF SCIENCE

Legislation to reduce the pollutants that cause acid rain has been opposed by a powerful coalition consisting chiefly of the midwestern electric utilities, the coal industry, and their supporters in the Reagan Administration and Congress. These opponents of acid rain controls have cited "scientific uncertainty" as the official explanation for inaction. "In light of the widespread uncertainty surrounding the causes and effects of acid precipitation," wrote the Edison Electric Institute, "additional emission controls on existing coal-fired power plants are not warranted at this time."[133] The U.S. Environmental Protection Agency and other agencies in the Reagan Administration supported the utilities' contention, arguing that "scientific uncertainties in the causes and effects of acid rain demand that we proceed cautiously and avoid premature action."[6]

This chapter examines the principal objections that have been invoked by the opponents of acid rain controls. Our primary concern, however, is not with science but with *method*. How was the acid rain debate shaped, and who shaped it? How was the public's understanding of acid rain influenced by the press, the scientific community, the Congress, industry, and the Environmental Protection Agency? Was the public interest well served? We will find that to an astonishing degree, the debate on acid rain has been marked not by science but by sophistry, by that ancient art of misleading without actually lying. The politics and science of acid rain became so thoroughly intertwined that they were barely distinguishable. Why was science politicized to such an unprecedented degree, what were the consequences, and what can be done to prevent a recurrence?

The scientific "Maginot Lines"

It should be emphasized that many aspects of acid rain are worthy of further research. Furthermore, reasonable people will disagree on the most equitable response to the problem. However, the basic facts about acid rain have been in hand for some time. At issue in this chapter is that the opponents of acid rain controls misrepresented or suppressed the basic facts of the problem — facts that were either clearly known to the parties involved or that reasonably should have been.

The principal objections raised by opponents of acid rain controls fall into the following categories:

"There is no acid rain problem."

"There may be a problem, but it has always been there."

"The problem is not getting worse."

"We don't know what causes the problem."

"The problem is due to natural causes."

"We may have caused the problem, but we don't need to act."

"We don't know how to act."

"Taking action won't do any good."

The validity of these objections is discussed below.

"There is no acid rain problem."

Acid rain was portrayed by the opponents of regulation as largely a problem of definition. "The term 'acid rain' is a redundancy," wrote Carolyn Curtis for the Electric Power Research Institute, the research arm of the electric utilities. *"By definition, natural rain is somewhat acidic . . .* So by all rights we should have been saying for years, 'It's acid raining outside' . . ." (emphasis original).[134]

Testifying before the Senate Energy Committee in 1982, the

E.P.A.'s then-chief of air pollution control, Kathleen Bennett, argued that "there is no good measure of when acidity in rain should be considered detrimental . . . Hence, at this point, there is no clear reference for developing a remedial program."[135] A former director of E.P.A.'s Office of International Activities, Richard Funkhouser, asked in an editorial in the *New York Times*, "Has science established precisely the dividing line between benign and malign acidification? All rain is naturally acidic, usually to the benefit of the environment . . ."[136]

Aside from the rhetorical device of asking for a "precise" dividing line (Is there a precise dividing line between "hot" and "cold"?), the E.P.A had already been given the guideline that it claimed to seek: In 1981, the National Academy of Sciences issued a widely-publicized report on the effects of acid rain which concluded that rainfall more acidic than pH 4.6 to 4.7 is harmful to sensitive regions, while less acidic rainfall is not.[38] (Rainfall in the Northeastern U.S. is currently two to five times more acidic than this guideline.) The Reagan Administration dismissed the Academy's report as "lacking in objectivity" but did not take issue with its scientific findings.

"The problem has always been there."

Industry portrayed acid rain as a natural, worldwide phenomenon that has been around for hundreds of years. Consider the following "news item" from the influential journal, *Environmental Science and Technology:*

> "*The notion that acid rain originates principally from electric power plants is being challenged by various scientists and organizations. Among items cited are literature reports of pH 4 rain in Paris in 1852 and evidence of about pH 5 rain in the Himalayas and Antarctica as long as 350 years ago. Moreover, National Oceanic and Atmospheric scientist John Miller cites low pH rainfall in northern Canada, Hawaii and Samoa — far away from coal-burning utilities. Other arguments are also set forth in studies sponsored by the Edison Electric Institute and others.*"[137]

On the face of it, this "news item" appears to absolve electric power plants from acid rain.[5] Closer scrutiny shows that the opposite is true:

• The "pH 4 rain in Paris in 1852" was known *at the time* to be caused by coal burning. (See Chapter 1.)

• The "pH 5 rain in the Himalayas" refers to a research report concerned not with rain but with ice cores from Himalayan glaciers: When glaciers form under enormous pressure they become supersaturated with carbon dioxide, much like soda pop. The mild acid this produces (carbonic acid) has nothing to do with the strong acids (sulfuric and nitric acids) that cause acid rain, a point that was made clear by the authors of the original research report.[138]

• John Miller did not suggest that power plants are unrelated to acid rain in the eastern U.S., as misleadingly implied in the item. On the contrary, Miller speculated that the acid rain he studied in Hawaii might originate with industrial pollution from Asia that is carried all the way across the Pacific.[89] Precipitation in Antarctica and Samoa was found in Miller's and other's studies to be *pristine*, not polluted.[61] (See data in Chapter 2.)

• Acid snow and haze as far north as the Arctic are known to be caused chiefly by coal-fired plants in Europe, Asia, and North America.[85]

By a combination of misstatement and innuendo, the "news item" leaves the reader with the false impression that acid rain is caused chiefly by something other than industrial pollution. While the first sentence of the item touts "various scientists and organizations," only in the last sentence do we learn that the source is the Edison Electric Institute and unnamed "others."

Writing in the Wall Street Journal, retired DuPont executive P.J. Wingate used the opposite approach to the same end: "Acid rain is no problem at all," he noted reassuringly, "on about 95% of the surface of the globe. The 95% estimate may startle some people who have been led to believe that acid rain is a problem nearly everywhere, or soon will be . . ."[139] A glance at a globe shows that 5% is hardly consolation, especially when that few percent happens to include eastern North America and western Europe.

"The problem is not getting worse."

The opponents of acid rain controls have argued that the acidity of rain is not getting worse — in fact may even be getting better — and that there is therefore no need to take action on acid rain soon. "Is acid rain increasing? We are not certain," wrote Ralph Perhac, the Director of Environmental Affairs for the Electric Power Research Institute. "Those who claim it is admit the changes are subtle on a yearly basis . . . We do not claim that the acidity of rain is not increasing. We claim only that the data are conflicting and do not allow one to draw firm conclusions."[140]

In the same vein, Richard Funkhouser of the E.P.A. wrote in *The New York Times*, "Do scientists have a firm idea whether acidification is getting better or worse? Swedish scientists, who have worked on this problem the longest, have been unable to discern a trend. Perhaps the best United States records were compiled between 1965 and 1975 in New York State by the United States Geographical [sic] Survey. Here too, no trend was identified."[136]

Both industry and the E.P.A. actually understated their case: The acidity of rain in some parts of the eastern U.S. has in fact *decreased* slightly in the past decade. The U.S. Geological Survey had found that the acidity of rain in New York State was either stable or decreasing,[141] and a similar long-term record at Hubbard Brook, New Hampshire also showed that acid sulfate levels in rain had decreased during the decade.[127] However, this decrease in acid sulfate was expected, because the Clean Air Act of 1970 required the major SO_2 emitters, many of them in the Midwest, to reduce emissions. (For industry or the E.P.A. to have emphasized this decrease would have been an implicit admission that reducing emissions really does reduce acid rain, a contention that the opponents fought adamantly at a later stage of the debate. See below.)

But the trend in the acidity of rain is a red herring. Acid rain is a *cumulative* problem; the effects are getting worse even though the acidity of rain in some parts of the East may be decreasing slightly. The reason is that many lakes and streams have lost so much of their buffering capacity that they are now exquisitely sensitive to additional acidity. (This effect is familiar to freshman chemistry students who have titrated a buffer with an acid; at the endpoint of the titration, even a small input of acid causes a large change in the acidity of the system. The naive observer sees a strange phenomenon at the endpoint: the indicator dye changes color more *rapidly* as acid is added more *slowly*!) At the current level of acid deposition, the number of acidified lakes is expected to double by 1990.[26] Only a sustained decrease in emissions will prevent this.

Furthermore, the decline in SO_2 emissions during the 1970s has been reversed. For one thing, some of the largest midwestern emitters of SO_2 are not in compliance with the Clean Air Act, and the Reagan Administration has sought changes that would further increase SO_2 and NO_x emissions. (See Chapter 1.) Despite provisions of the Clean Air Act that strongly limit SO_2 emissions from new power plants, the utilities themselves project increases in SO_2 emissions of about 4 million tonnes a year (20%) by 1990.[142]

"We don't know what causes the problem."

The energy industry claimed for a remarkably long time that acid rain in the eastern U.S. is caused by "natural factors," rather than fossil fuel combustion. They succeeded in this deception largely because the public rarely asked for *numbers*. Three numbers would have sufficed to settle the issue: How much sulfur does industry put into the air? How much sulfur does nature put into the air? How much sulfur returns to earth in acid rain and other forms of acid deposition? These numbers not only were known for quite some time, they came from inventories developed by industry and the Environmental Protection Agency themselves.[64] As discussed in Chapter 3, about 30.6 million tonnes of sulfur (calculated as sulfate) are emitted annually by man-made sources, compared with only 0.36 million tonnes by the leading natural source, biological emissions. Thus more than 95% of sulfur emissions in the eastern U.S. are man-made. These man-made emissions account not only for the 7.3 million tonnes of acid sulfate that fall on the eastern half of the U.S. in acid precipitation, they also account for all the sulfur pollution that falls in dry form, as well as the sulfur pollution carried downwind to Canada and out over the Atlantic Ocean. (See Table 4.) Similar conclusions apply to nitrogen emissions.

Despite this unambiguous record, industry continued to claim that "natural sources" of SO_2 and NO_x are a significant part of the acid rain problem and that industry's contribution is "uncertain." Standard Oil of Ohio suggested in its Summer 1982 Newsletter, *Energy and You*, "It's not clear how great a factor nature itself is in acid rain. Significant amounts of acid-forming compounds are released by volcanoes, lightning and forest fires." Continuing the theme, the E.P.A.'s Richard Funkhouser wrote in *The New York Times* in 1983, "Does the state of science tell us what amount of acidification is manmade and what amount occurs naturally? . . . The eruption of one Mexican volcano, Chichonal, reportedly threw more sulfur into the atmosphere than a dozen Ohio River Valley industrial complexes."[136]

Yet El Chichon volcano contributes negligibly to acid rain in the eastern U.S. — just as the radioactive fallout on Ohio from a remote nuclear test would not be comparable to that from a bomb dropped on Cleveland. El Chichon hurled an estimated 20 million tonnes of acid sulfate into the stratosphere — more than 15 miles high — where the acid circled the globe and remained for more than a year, becoming highly diluted in its travels. The eruption was described as "a once-in-a-century event that has already produced a 3° to 5° C. heating in the stratosphere and will probably cause a measurable climatic perturbation . . . The El Chichon eruption of 1982 is the largest such event in terms of lasting effects on the stratosphere . . . and probably will prove to be the largest in at least the past century."[68] The Ohio Valley polluters hardly seem benign in comparison. They emit a comparable amount of sulfur, year in and year out, and they inject it not into the stratosphere but into the lower layer of the atmosphere in which we make our home.

The news media perpetuated the myth that natural factors cause acid rain in the eastern U.S. Consider what happened in 1982, when a *New York Times* reporter visited the laboratory of Dr. Volker Mohnen, director of the Atmospheric Sciences Research Laboratory at the State University of New York at Albany. Unknown to the reporter, Dr. Mohnen had been retained by the nation's largest coal company, Peabody Coal, to testify on the company's behalf at an E.P.A. hearing on interstate air pollution. (For a discussion of the E.P.A. hearing, see Chapter 8.) At that hearing, Dr. Mohnen argued strongly against reducing SO_2 emissions from midwestern power plants.[143] As we shall see, Dr. Mohnen subsequently played a key role in the Congressional debate on acid rain as the chief scientific proponent for the industry's viewpoint, although his industry ties were seldom made explicit.[144]

"Dr. Mohnen sought to clear up misconceptions for visitors to the station . . . ," the *Times* reporter wrote. "Dr. Mohnen said, while the acid rain problem in the U.S. and Canada is widely assumed to be caused by industrial sources such as coal-using power plants, much of the acidity is due to natural causes such as rotting vegetation and the emissions of sulfates by seawater."[145] According to the comprehensive study funded by the utilities themselves, "rotting vegetation" and other natural sources contribute less than 5% of the sulfur emitted in the eastern U.S. (See

Chapter 3.) The reporter had neglected to ask *how much* acid rain is due to "natural causes," and was thus carefully left with the wrong idea.

Throughout the acid rain debate, opponents of acid rain controls placed the burden of proof on the scientific community, in accord with the lawyer's maxim, "The party with the burden of proof will lose." An interesting example was the charge that scientists have not *proved* that acid rain has caused lakes to acidify in the Northeast. In March 1983, for example, *The New York Times* ran the headline, "Acidity in Lakes Attributed to Natural Chemicals."[146] The article referred to the claim of a consulting geologist, Dr. A. Gordon Everett, who was hired by Consolidated Edison Company to testify concerning the Company's application to burn coal instead of oil in three of its New York City generating plants. As *The Times* put it, "Dr. Everett said his studies had convinced him that while there *might* be acid coming down in rainfall, the predominant cause of acidity in the Adirondack waters was the naturally acid materials in the region being flushed into lakes by the rain" (emphasis added). The "naturally acid materials" were attributed to "abundant decaying vegetation."

Of interest here is not Dr. Everett's contention, which has been discussed elsewhere, but the sleight-of-hand that could induce the reporter to write that "there might be acid coming down in rainfall." Throughout the eastern U.S., rainfall is now far more acidic than any of the lakes and streams it replenishes. In fact, if lakes and streams ever became as acidic as rainwater, they would be so acidic that every body of water east of the Mississippi would be virtually devoid of life. *Prima facie*, the question is not, "Why have lakes acidified?," but rather, "What is preventing lakes from becoming even *more* acidic than they are now?" The answer is that for most lakes, the surrounding rocks and soils partially neutralize acid rain; only in those watersheds incapable of tempering acid rain have lakes acidified. From this perspective, the burden might reasonably be on the advocates of inaction to convince the public that watersheds throughout the U.S. are indeed adequate to buffer acid rain and that in the future, not all lakes will become as acidic as the rain that falls on them!

"We don't know which polluters are responsible for acid rain."

No aspect of acid rain has been more bitterly contested by industry than the question of which polluters are responsible for acid rain in the Northeast and other sensitive regions. The evidence shows that while the Northeast contributes to its own acid rain, the majority of its acid rain comes from upwind sources in the Ohio Valley and Midwest. (See Chapter 4.)

The energy industry succeeded in shifting attention away from the actual sources of pollution. The news media reported that scientists were unable to "pinpoint" the sources of acid rain or to "trace a given molecule of pollution back to its source." ("Pinpoint" was a curiously incongruous term to describe smokestacks that are skyscrapers and that emit as much SO_2 annually as Mt. St. Helens volcano. And imagine walking into room filled with cigarette smokers and asking to "pinpoint" the sources of smoke by tracing each parcel of smoke back to its source!)

In reality, the street address of every major SO_2 emitter in the eastern U.S. is on file with the E.P.A. There is no mystery about their location. The sphere of influence of these SO_2 emitters has been estimated by many research groups using computer calculations that incorporate such factors as wind speed, wind direction, smokestack height, etc. No one has pretended to be able to calculate the fate of every molecule from every smokestack; nevertheless, computer calculations provide a quantitative estimate of the amount of sulfur pollution each source contributes to any given location downwind. In the case of the Adirondacks, for example, there is remarkably good agreement between computer calculations and actual field measurements: Both techniques indicate that greater than 50% of the acid pollution in the Adirondacks comes from sources in the Ohio Valley and Midwest sector, while less than 10% comes from the Northeast itself. (See Chapter 4 and Tables 5 and 6.)

Again, the burden of evidence might reasonably rest with the polluters themselves. Of the many hundreds of power plants in the eastern half of the U.S., just the 20 largest coal-fired plants are responsible for about 20 to 25% of all airborne sulfur. That is an amount of sulfur that *in principle* could account for all the acid sulfate falling in acid precipitation on all states east of the Mississippi combined (cf. Tables 2 and 3). One is entitled to ask, Where does all that sulfur *go*? As one physicist put it, "They'd have you believe that when the wind blows, it doesn't carry the sulfur with it!" Despite the record, the midwestern utilities denied that they contribute heavily to the Northeast's acid rain and even attempted to shift the blame onto the Northeast itself. They did it largely by referring to the claims of just two scientists: Dr. Volker Mohnen (mentioned above), and Dr. Kenneth Rahn, an atmospheric chemist from the University of Rhode Island. It is not the purpose of this discussion to question the credibility of one or another individual, but rather to show that the so-called "disagreements within the scientific community" often boiled down to the minority views of one or two scientists. The extent to which these two men influenced the course of the acid rain debate on the basis of evidence that was neither published, nor peer-reviewed, nor substantially correct is one of the truly remarkable episodes in recent science policy.

Consider the field study on the sources of acid rain in the Adirondacks (discussed in Chapter 4). The study was sponsored by the utilities and Dr. Mohnen had played a prominent role. The results hardly gave comfort to the midwestern utilities: 62% of the acid sulfate deposited on the Adirondacks in precipitation was found to come from the Ohio Valley/Midwest, while less than 5% came from the Northeast itself. (The reason for the twelve-fold disparity is that precipitation coming to the Adirondacks from the Midwest was *four* times more polluted and *three* times more plentiful than precipitation coming from the Northeast. See Table 6.) The results indicated clearly that in order to significantly reduce acid rain in the Adirondacks, one would have reduce emissions in the Ohio Valley / Midwest. Even eliminating *all* of the Northeast's emissions would have an insignificant effect on acid rain in the Adirondacks.

Consider how easily these results were circumvented. In May 1981, Dr. Mohnen referred to the study when he testified for the Peabody Coal Company at E.P.A. hearings on interstate pollution. "While one might expect to see significant differences in precipitation chemistry as air masses from different regions arrive at (the Adirondack) site," he testified, "the analysis showed . . . that the *precipitation volume,* more than any other factor, determines the amount of deposition for the three pollution-related ions (acid, sulfate, and nitrate)."[143] That is, acid rain depends primarily on the fact that it rains! Scientifically, that was gibberish, but it was clever enough to shift attention away from the pollutants and onto the rain itself. Dr. Mohnen's testimony gives the impression that there *aren't* "significant differences" in precipitation chemistry — that rain is just as polluted no matter where it comes from. But the study showed that midwestern precipitation is *several times* more polluted with acid, sulfate, and nitrate than northeastern precipitation. (See Fig. 5) In his testimony, Dr. Mohnen omitted the data for the Northeast sector; this discouraged comparison between the Northeast and Midwest sectors.

The news media disseminated the myth that the sources of acid rain are mysterious. National Public Radio interviewed Dr. Mohnen and announced that "At least one scientist says that the Ohio Valley is being blamed prematurely as the source of acid rain in Canada and New England."[147] In the ensuing interview, Dr. Mohnen again referred to the Adirondack study: "We ask, where has the air been over the past two days that brought (to the Adirondacks) this pollution in the rain? We find that *it does not matter where the air has been.* Simply, there is always enough material in the air to be converted into acidity that comes down with the rain." Again, this gives the false impression that rain is equally acidic "no matter where" it comes from. Mohnen concluded, "That is the major result that we have to keep in mind: If we live in Albany, N.Y., New York State must look at New York State *first*, because there are sulfur emissions that contribute to acid rain." Again, Dr. Mohnen neglected to mention *how much* of the Adirondacks' acid pollution actually comes from New York State. Considering that the study to which he referred showed that 62% of the acid sulfate in Adirondacks precipitation comes from the Midwest sector — more than twelve times as much as from the Northeast — the Ohio Valley was hardly being blamed "prematurely" as the source of New England's acid rain!

Dr. Mohnen continued to blame the rain itself for acid rain when he testified before a House Subcommittee in late June 1981, but when he tried the same argument before the Senate Energy Committee in 1982, he ran into trouble from Senator George Mitchell.[148]

"Dr. Mohnen . . . as I understand your conclusion, it is that the quantity of sulfate . . . in rain that came from the Midwest sector was roughly comparable to that which came from other sectors. Is that a fair summary?" Mohnen balked. He had to acknowledge that two thirds of the acid sulfate in rain came from the Midwestern sector. He was then forced to address the finding that so little of the pollution came from the Northeast.

Dr. MOHNEN. There is another sector that I personally disqualified after talking to many of my colleagues, simply because when we talk about air from the Northeast sector or from the East, this is mostly a northeastern snowstorm, and you know what that is. So obviously you wouldn't want to compare material that was deposited by a Northeastern snowstorm with something that is deposited by the more typical precipitation patterns that come (from the) Northwest-Southwest.

Sen. MITCHELL. But what I am getting at is . . . what you are saying is that the quantity of acid deposition was dictated by the amount of rain that fell from a given sector.

Dr. MOHNEN. On an annualized basis, yes.

Sen. MITCHELL. So there is more rain coming from the Midwest sector; therefore, there is more acid coming from the Midwest sector.

Dr. MOHNEN. That is also correct.

Sen. MITCHELL. So there are two variables in there. One is the quantity of rain and one is the quantity of acid.

Dr. MOHNEN. Right.

Sen. MITCHELL. Now we can't do anything about the rain, can we?

Dr. MOHNEN. Totally correct.

Sen. MITCHELL. We *can* do something about the quantity of acid.

* * *

In 1982, the National Academy of Sciences asked Dr. Mohnen to join an eight-member panel whose mission was to summarize the scientific evidence on how acid rain is formed and transported through the atmosphere. The panel's deliberations bore on two important questions: Which sources of pollution are responsible for acid rain in the Northeast and elsewhere in the eastern U.S., and would reducing emissions commensurately reduce acid rain? Stung by the Reagan Administration's criticism of its earlier report on acid rain, the Academy was apparently determined to bend over backwards to prove its "objectivity." The Academy enlisted Dr. Mohnen even though at the time, he was still testifying on behalf of Peabody Coal before the E.P.A.[149] (Indeed, another panel member, George Hidy, had also testified for Peabody at the E.P.A. hearings.) This possible conflict of interest was not noted in the biographical sketches accompanying the panel's report.

The Academy's report[63] was to have been a quick one but it was not released until June 1983. A section discussing the sources of acid rain was written in part by Dr. Mohnen. The report dismissed out of hand computer calculations of transport, work that had been done by many research groups, including those reviewed by the joint U.S./Canadian team set up under the Memorandum of Intent. "(Computer) models have pointed to the importance of certain geographical groupings of SO_2 sources and the potential influence of the sources on certain receptor areas," the report said cryptically. "However, current models have not provided results that give confidence in their ability to translate SO_2 emissions from specific sources or localized groupings of sources to specific sensitive receptors." On page after page, the message was the same: "Researchers in the field generally have only limited confidence in current results . . . We do not regard currently available models as sufficiently developed to be used with confidence . . . Their accuracy is inadequate for quantitative assessment of the effects of emissions from specific sources . . . We advise caution in projecting changes in deposition patterns that result from changes in emissions . . . Without greater confidence in the results of (computer) models, we cannot judge the consequence of emissions reduction in . . . the Midwest, for deposition in another region, such as the Adirondacks." Yet nowhere did the report actually present the results of *any* computer models, let alone evidence of their inaccuracy.

In a section of the report describing field observations on the sources of acid rain, Mohnen cited the Adirondacks study, but again circumvented its findings. "The evidence . . . suggests that the industrialized region to the southwest is a major source of both nitric and sulfuric acids deposited (in the Adirondacks). *However, the high percentage of acid deposition in the Northeast may be a result of the high percentage of total precipitation that is delivered by these parcels"* (emphasis added). Again, this was gibberish — if it didn't rain, there would *be* no acid rain — but it diverted attention from the pollutants themselves. Mohnen concluded, "On the basis of currently available empirical data, we cannot in general determine the relative importance for the net deposition of acids in specific locations of long-range transport

from distant sources or more direct influences of local sources."
This is known charitably as a whitewash.

At first, the press ignored Mohnen's conclusion in favor of the
major conclusion of the report, written by the panel's chairman,
Jack Calvert — a scientist at the National Center for Atmospheric Research — that reducing SO_2 emissions would reduce
acid rain commensurately. Dr. Calvert's conclusion was something of a rebuff to Dr. Mohnen and to the E.P.A. (In Senate testimony in 1982, Mohnen had speculated that reducing emissions
by 50% "might not" reduce acid rain at all, a claim echoed by the
E.P.A.'s representative at the hearings, Dr. Kenneth Demerjian,
as a reason for going slow on acid rain. See Chapter 6 for a discussion of the science.) The report's conclusion was widely reported
in the press with such headlines as "Acid Rain Tied Directly to
Emissions." As Calvert later put it, if industry "gets off the dime"
and reduces emissions, "we'll guarantee an effect."[150] But
Mohnen then sent letters to *The New York Times* and *Time*, ex-
pressing concern that the report was being "misinterpreted",
adding that "the contribution of midwestern sources to acid rain
in the Northeast remains unknown."[157]

The Academy's imprimatur on the report lent Mohnen's conclusions a patina of scientific credibility to which the opponents
of acid rain controls could refer. In an editorial in *The New York
Times* in August 1983, Sen. Richard Lugar of Indiana cited the
report. The Senator wrote, "The widely publicized conclusion of
the Academy report is that reductions in sulfur and nitrogen oxides will bring about corresponding reductions in acid rain. Less
well reported, however, are two other conclusions that, I believe,
could break the legislative stalemate that now exists on the acid
rain issue . . . It is stated in the report that 'we cannot judge the
consequences of emissions from a region such as the Midwest,
for deposition in another region, such as the Adirondacks or
southern Ontario.' The second conclusion is that the closer a
source of emissions is to an affected area, the greater the benefits
that can be achieved from a given reduction in emissions. The

report is thus as much an indictment of proposed acid rain legislation as it is a call for action."[152] Sen. Lugar concluded that "the
cost of the cleanup can be lowered by targeting reductions as
close as possible to the affected areas of the Northeast."

Thus was responsibility for the Northeast's acid rain shifted
from the Midwest to the Northeast.

* * *

The debate about which polluters are responsible for the Northeast's acid rain took a remarkable turn on the basis of a letter sent
to members of Congress in November 1981, by a researcher from
Rhode Island named Kenneth Rahn.[153] Dr. Rahn explained that
he had "quite accidentally produced some new data" — all of it
unpublished — which he believed cast doubt on the view that
the Midwest caused much of the Northeast's pollution.

Rahn's work was based on the fact that different pollution
sources emit characteristic chemical "signatures" — characteristic amounts and types of chemical elements — depending on
the type of fuel burned. For example, coal-fired plants emit more
manganese than vanadium; oil-fired plants emit more vanadium
than manganese. By measuring the levels of these chemical elements in air pollution, Rahn hoped to determine whether the
pollution came from coal-fired sources (and thus predominantly
from the Midwest), or from oil-fired sources (and thus chiefly
from the Northeast). The method is highly indirect.

Rahn concluded in his letter, "Instead of a monolithic midwestern source [of pollution], the Northeast is now seen to have
a rich variety of sources and transport, a complexity that was
totally unanticipated; the Sudbury plume [from the INCO nickel
smelter in Ontario] can be detected after transport across the
entire Northeast; the Southeast pollutes New England; the East
Coast pollutes itself considerably. Ironically, the (midwestern)
source that we had expected to find the most easily has proven
the hardest to detect." Although he warned that "perhaps some of
our results depend on the nature of the technique . . . perhaps

not," Rahn urged Congress to "learn with *certainty* which sources are most important *before* attempting massive remedial action" and he proposed a "systematic study" using his technique (emphasis added). It was unusual for a scientist to make such a direct political appeal on the basis of data and methods that were neither published, nor peer-reviewed, nor made generally available to researchers in the field.

Rahn's work was lent apparent credibility when it was featured in the Research News column of *Science* — a column usually devoted to fast-breaking but substantiated research news — with the headline, "Tracing Sources of Acid Rain Causes Big Stir."[154] The subheading read, "New data suggest that the Midwest may not be responsible for all of the Northeast's acid rain."

As Rahn's technique was more closely scrutinized, however, it began to fall apart. George Wolff, a scientist from General Motors, pointed out in a letter to *Science* that Rahn's work made "two major assumptions that are questionable."[77] One assumption was that the "source areas of manganese and vanadium are the same as the sources of acid precipitation." Wolff noted, for example, that the New York metropolitan area is the "largest source of vanadium in the Northeast." Thus air from the Midwest that passes over New York will pick up vanadium and will seem to have a local "signature", even though the bulk of the sulfur in the air mass may have come from the Midwest. Wolff concluded that because of this and other faulty assumptions, "the technique used by Rahn does not appear to be valid." He noted that "Rahn's hypothesis concerning the local contribution of (acid) sulfate aerosol is contrary to observations and (computer) modeling results that demonstrate the importance of long range transport of (acid) sulfate."

When Rahn appeared before the Senate Environment Committee in June 1982, he was confronted by Senator Mitchell. The senator pointed out that the chemical elements Rahn had chosen to use as tracers were emitted as coarse particles that settle to earth relatively quickly, while sulfur pollution can travel further downwind before returning to earth. Thus if relatively little manganese reaches the Northeast, it may give the false impression that little sulfur is reaching the region as well.

Rahn retreated, saying that "the statement of last November which has caused such a fuss is a result of spare time and effort. It is the result of not large amounts of effort on our part."[155] Mitchell was incensed.

Sen. MITCHELL. I find it incredible that on the basis of admitted part-time, spare-time study, as a spinoff of something else, in an area that you had no previous experience with, that involves 10 percent of the problem, you come in here and make a very emphatic and specific recommendation, "Here is what we should do. Here is what we shouldn't do."

Rahn retreated even further. "I *believe* in long-range transport . . . I believe it is more common and widespread than is recognized today . . . But is that all there is to this story in the Northeast? That is what I have set about to try to answer here."

Sen. MITCHELL. Has anybody, to your knowledge, ever suggested that all of the deposition in the Northeast or East comes from the Midwest?

Dr. RAHN. Practically all of the popularized accounts, the thing that is in the heads of most Americans in the Northeast, I believe, rightly or wrongly so, is that the Midwest is like a monolith. It spreads its aerosol over the Northeast and —

Sen. MITCHELL. What do you base that on?

Dr. RAHN. I base that on many, many, many statements in the media, in the popular press, on basically listening to everyone else.

Sen. MITCHELL. I just want to say, Dr. Rahn, I have discussed this subject hundreds of times in the East — hundreds of times before literally thousands of people, and I have never heard anybody make the statement that you now say everybody in the East believes, that it all comes from the Midwest. We have been very careful, those of us who have been proponents, to make clear that this is a regional problem and the sources are region-wide

. . . Both you and Dr. Mohnen have repeatedly used the phrase of making one area do something to the benefit of another . . . Have you read my bill, S. 1706?

Dr. RAHN. Parts of it.

Sen. MITCHELL. Parts of it?

Dr. RAHN. Parts of it.

Sen. MITCHELL. But you come in here and recommend strongly against a bill you haven't even read in full; is that right?

Dr. RAHN. That is correct.

As an illustration of the supposed power of his tracer technique, Rahn had claimed to be able to detect "the plume from the Sudbury, Ontario INCO smelter which emits 10% of the North American SO_2." However, scientists from New York State pointed out that Rahn could not possibly have detected pollution from the INCO nickel smelter because, during the period Rahn's measurements were made, the smelter had been shut down by a strike!

None of this stopped the E.P.A. from seizing upon Rahn's work as an excuse for calling legislation on acid rain "premature." Appearing before the Senate Energy Committee in August 1982, the E.P.A.'s then-chief of air pollution control, Kathleen Bennett, testified, "The uncertainty surrounding acid deposition is demonstrated by the divergent conclusions reached in two recent studies. One study [Rahn's], conducted by the University of Rhode Island . . . indicates that acid deposition in the Northeast may result from sulfur dioxide emissions of local origin rather than from Midwestern sources . . . Another recent study by the Mitre Corp. concludes that the current acid deposition problem is largely a long-range transport problem . . . [These] opposing theories on the cause of the acid deposition problem emphasize the degree of the current uncertainty . . . and the consequent need for further research."

The E.P.A. went further: it shifted responsibility for acid rain to the Northeast. As mentioned above, an interim report from the U.S./Canada joint study had presented computer calculations showing that the majority of acid sulfate in the Adirondacks came from the Ohio Valley, Midwest, and Tennessee Valley (Table 5). But when the E.P.A. released its Final Report in late 1982, these results were not to be found. Instead, the report focused on the hypothetical influence of *proximity*, not on the actual *amount* of pollution emitted by real sources. This ploy enabled the E.P.A. to conclude that the Northeast (New York to Maine) "has the greatest impact" on the Adirondacks, while the Ohio Valley (Pennsylvania, Ohio, West Virginia) and Ontario "have the second greatest impact at the site."[156] To the casual observer, this suggests that the Northeast is most responsible for acid rain in the Adirondacks. But the conclusion means only that a hypothetical power plant would do more damage if were sited in the Adirondacks itself than if it were sited hundreds of miles away in Ohio. In reality, the Midwest and Ontario, even though they are far from the Adirondacks, emit so much sulfur that they contribute the majority of acid rain to the Adirondacks. Telling the Northeast to reduce its emissions first would be like telling a homeowner to fix a leaky faucet while flood-waters were lapping at his door.

The scientific press seemed confused by all this. In an editorial on acid rain in July 1983, *Science* asserted that " . . . there is wide disagreement among sincere people as to . . . who is responsible, and how the problem should be ameliorated . . . People in the northeast United States take the position that coal-fired utility plants in the Midwest are a principal source of the acid in the rain that has been falling on them . . . A large number of studies, however, have shown that the Northeast is itself responsible for a large share of its own pollution. Indeed, everyone who drives an automobile is a contributor to acid rain."[157]

Commendably evenhanded, but was it true? Where *were* the "large number of studies" and how "large" a share of its own acid rain was the Northeast claimed to produce? The answers were not to be found in the pages of *Science*. What was to be found — in as respected a journal as *Science* and in much of the press —

were not *numbers*, but opinions, hearsay, and quotes taken out of context from reports that few seemed to have read. For example, a report from the Congressional Office of Technology Assessment was widely quoted as concluding that the Northeast "contributes as much or more to its own (sulfur) deposition as any other single region contributes to it."[27] This conclusion gives the impression that the Northeast should control its own pollution first — an impression that was not lost on Congressional opponents of acid rain controls. But closer inspection show that the O.T.A. report defined the "Northeast" to include the state of Pennsylvania — a state which emits more sulfur than all Northeastern states combined; and the "other regions" to which the Northeast was compared were chosen to include clusters of just a few states. This ensured that no other single region would contribute more to the Northeast than the Northeast itself.

As the *Science* editorial pointed out, everyone in the Northeast who drives an automobile produces NO_x emissions which contribute to acid rain. But how much do they contribute? The majority of NO_x emissions are known to originate from smokestacks, not autos. Indeed in the Adirondacks study most of the acid nitrate falling on the Adirondacks was found to come from the Midwest — only 7% came from the Northeast itself (Table 6).

In short, the scientific record — imperfect though it is — shows clearly that in order to significantly reduce acid rain in the Northeast it is necessary to control the major sources of pollution in the Ohio Valley and Midwest. The opponents of acid rain controls largely succeeded shifting the blame to the Northeast itself.

Honesty is the best policy, but not the only policy

The most disturbing part of the acid rain debate is the ease with which fact and fiction became partners. Science and sophistry, earnestness and artifice, were seated as equals at the same tables of power. We are not talking here of truth, but of truthfulness. Any scientist worth his salt is *wrong* ten times a day if he is at the cutting edge of his field. But science carries with it an ethic that is perhaps unique in all fields of endeavor: it is forbidden to mislead. That is an ethic in increasingly short supply in a world in which image is all and facts are mere convenience. In the hands of those who would abuse it, science became less a tool for solving the nation's problems and more a means to a political end, a mere debating tactic.

The policy of making science the scapegoat for inaction had three serious effects. It squandered both time and money that could have been better spent in mitigating acid rain. It set a dangerous precedent for the political abuse of science. And it eroded the credibility of the Environmental Protection Agency and other agencies in the Reagan Administration that rely on science in formulating public policy. In the thermonuclear age, in which we literally entrust the government with our lives, ought not credibility be taken seriously?

How can we ensure that in the future, public policy is better served by science? Here, are several suggestions:

First, it is unwise to entrust the same government agency — in this case the Environmental Protection Agency — with responsibility for funding research and also for setting and enforcing public policy. This conflict of interest had a chilling effect on the scientific community. To put it bluntly, few scientists were willing to jeopardize their research funds by publicly criticizing the E.P.A's interpretation of the scientific record.

Second, we must ask of those who invoke "uncertainty" as a reason for inaction, Specifically *what* is uncertain, and *how* uncertain is it? We must ask for numbers, for estimates, for upper and lower limits of error. Those who seek scientific "certainty" are pursuing an illusory goal — or are stalling. In the real world, every measurement is subject to further refinement; almost all public policy is based on information that is "uncertain" yet within a suitable margin of error. As Senator Mitchell put it, "The argument that we have to wait until we have the complete answer to every part of the problem is the argument most used by people who want inaction, want no action. And were we to wait for that day, it will be your grandson and my grandson who will be discussing the issue then. Human affairs do not lend themselves, and never have lent themselves, to the approach that you have to have a precise scientific answer to every conceivable question that exists before you take any legislative action."

Third, it is dangerous to allow scientific debates to be waged as an arcane war between "experts." We must be willing to examine the evidence for ourselves. Admittedly, that is a tall order. It places tremendous demands on the public's understanding of science and on scientists' responsibility to communicate with the public clearly and accurately. But it is a goal to which we must aspire if we are not to be manipulated.

Finally, we need an adequate forum for resolving scientific disputes that have a strong political component. The forum should have as its most important feature a mechanism for *converging* on accurate science; it should encourage the dialectic that is at the core of science. Existing forums were clearly not up to the task. The Congressional hearings on acid rain were governed largely by political considerations and rarely by the norms of scientific discourse. The E.P.A.'s hearings on interstate pollution allowed no cross-examination of witnesses, and the voluminous record generated by the hearings was only for the hardy to assimilate. Scientific reports sprang up like mushrooms after a rain — reports that, in the words of one Washington bureaucrat, were "of a weight you wouldn't want to send first-class mail. You wouldn't want to *read* them . . ." Even the U.S. / Canada bilateral study could not achieve consensus on the science of acid rain. "Science by publicity" became the strategy of choice: electric utilities mailed their customers millions of misleading pamphlets on acid rain; environmental groups traded charges and countercharges; hastily assembled interest groups took out expensive ads on editorial pages; scientists lobbied Congressmen on the basis of unpublished and inaccurate results. Surely there is a better way for science to inform public policy.

It is worth scrutinizing the words of Carl Bagge, Director of the National Coal Association, from a speech on acid rain: "The central issue for us today is this: What kind of a society are we forging for the future? I submit that the parapolitical forces opposed to economic growth, those who find acceptance masquerading as environmentalists, have a vision of our future that is not shared by most Americans. It is up to us to stop them from defrauding the American public, whom they manipulate and do not trust."[50]

Fraud, manipulation, and trust are indeed the issues. The growing abuse of science for political ends is casting a shadow over the credibility of our leaders and unless we stop it, we will all be chilled by the darkness.

8

ACID RAIN AND THE LAW

Four words inscribed on the Supreme Court Building in Washington, D.C., go to the heart of the acid rain dispute: "Equal justice under law." At issue is whether it is fair for the air pollution from one state to interfere with the air quality and environmental health of a downwind state. The northeastern states have not waited for the Congressional storm over acid rain to resolve itself; they have gone to the heart of the issue by invoking existing laws governing interstate transport of pollution. This chapter describes a little publicized but important hearing on interstate pollution involving the northeastern states, the Environmental Protection Agency, and the midwestern utilities. It provides a glimpse of how environmental laws are administered, and illustrates how easily the law can be manipulated for political ends.

Acid rain and the Clean Air Act

The Clean Air Act is the public's first line of defense against polluted air. Hammered out in a series of bitter Congressional debates in 1970 and amended in 1977, the Act has attracted the wrath of both industry and environmentalists for either going too far or not going far enough on air pollution; yet is widely regarded as perhaps the most important environmental legislation of recent times. On the face of it, the Clean Air Act appears to encompass acid rain: It empowers the Environmental Protection Agency to regulate both SO_2 and NO_x emissions, and it also has a provision for dealing with interstate pollution. As we shall

114

see, however, the wording of the Act is subject to differing inter-
pretations and permits ample room for inaction.

The Clean Air Act sets limits on the airborne levels of several
air pollutants, including the SO_2 and NO_x which cause acid rain.
These limits are called "air quality standards." Each state must
submit to the head of the E.P.A. a plan for ensuring that emissions within the state will not violate the air quality standards
specified by the Act. Furthermore, no state may allow its emissions to "significantly interfere with the attainment and main-
tenance" of air quality standards in another state, according to
Section 110 of the Act. For example, SO_2 emissions from a power
plant in Ohio cannot be allowed to interfere with the air quality
standards for SO_2 across the border in Pennsylvania. In the event
of a dispute among states concerning air pollution, the E.P.A. is
required to hold a formal hearing to determine whether the provision on interstate pollution was in fact violated.

A major test of the Clean Air Act came in 1980, when New
York State used the provisions governing interstate transport of
pollution to attempt to hold the line on SO_2 emissions in the
Midwest. The focus of the case was a 1980 ruling by the E.P.A.
that allowed several midwestern states to *increase* their SO_2
emissions by a total of nearly two million tonnes a year. The
Agency reasoned that the increased pollution would not unduly
affect local air quality in the Midwest. But New York State feared
that much of the pollution would be carried downwind to the
Adirondacks and other parts of the state, would interfere with
the state's air quality, and would aggravate the state's already
severe acid rain problem. Accordingly, New York's Attorney
General, Robert Abrams, petitioned the E.P.A. to revoke ap-
proval for the increased pollution.[158]

The E.P.A. granted New York a hearing in the spring of 1981,
but refused to allow testimony on acid rain. The Agency argued
that it has no legal authority to consider acid rain, because the
Clean Air Act only sets limits on SO_2 but not acidity or sulfate.

As soon as SO_2 emissions are transformed into acid sulfate — a process that takes from several hours to several days — the pollution is beyond the law and out of the Agency's regulatory reach. "For these reasons," the Agency ruled, "E.P.A. requests that the parties presenting oral or written comments at the forthcoming public hearing refrain from basing their arguments on the cause and effects of acid deposition . . ."[159]

New York was forced to take an indirect approach. While there is no air quality standard for acid sulfate, there *is* an air quality standard limiting the amount of particulate matter in the air, and acid sulfate haze is in fact a major component of airborne particulate matter in the Northeast. New York presented evidence that the Midwest's SO_2 emissions contribute heavily to acid sulfate haze in the Northeast and "significantly interfere" with New York's air quality, on occasion even violating the federal limits on airborne particulates.

The utilities and the coal companies argued that scientific uncertainties precluded the E.P.A. from ruling in favor of New York. The legal counsel for more than a dozen of the midwestern utilities, Roger Strelow, summed it up this way: "Rather than pursuing baseless accusations aimed at an arbitrarily selected group of sources [the midwestern utilities] whose only vice is location outside of petitioners' jurisdictions, E.P.A. should turn this premature finger-pointing exercise into a constructive effort to develop adequate information. Petitioners, meanwhile should put their own house in order instead of trying to cast blame elsewhere for essentially local pollution problems."[160] (In what would appear to be an odd conflict of interest, Mr. Strelow was chosen by New York State in 1982 to be chairman of the Section on Economic Incentives at a statewide conference on acid rain, hosted by New York's Governor Hugh Carey and the Department of Environmental Conservation.[161] By some oversight, the Governor neglected to invite the scientists and lawyers from New York's own Department of Law — the very people who were pressing New York's case before the E.P.A. and who had amassed an impressive amount of evidence on the origins and effects of acid rain. According to one official, Governor Carey had presidential ambitions and sought to win favor in the Midwest by going easy on acid rain. Another explanation is that New York's energy agency was pressing for the conversion of oil-fired power plants to coal, and was only too happy to downplay the link between coal-fired plants and acid rain.)

Attorneys for the utilities and coal industry exposed other apparent loopholes in the Clean Air Act. For example, nowhere does the Act specify exactly *how much* air pollution must be transported across state borders to constitute "significant interference". This determination is apparently left to the discretion of the E.P.A. administrator. The utilities also argued that the Clean Air Act forbids the E.P.A. from taking into account the *aggregate* effect of midwestern pollution. Rather, they argued, each *individual* power plant must be proven responsible for violating the limits on airborne pollution hundreds of miles downwind. That interpretation of the Clean Air Act would present a Catch-22, a "fallacy of the commons," by which all polluters would escape regulation unless responsibility could be assigned to each polluter individually. In rebuttal, New York argued that on the contrary, the Clean Air Act requires the E.P.A. to consider the aggregate effect of midwestern emissions.

The E.P.A opposed attempts to close some of these loopholes. Sen. Dodd (D., Conn.) introduced a bill that would have allowed the E.P.A. to take acid rain into account in considering New York's petition, only to have the bill opposed by the E.P.A.'s then-chief of air pollution control, Kathleen Bennett. She argued that Sen. Dodd's bill would grant the E.P.A. administrator "broad discretion to require potentially costly emission reductions for pollution control" without the guidance of a specific air quality standard for acid rain.[6]

Yet the E.P.A. urged Congress *not* to draw up a specific air quality standard for acid rain, on the grounds that the existing Act was adequate. Appearing before the Senate Energy Commit-

tee in 1982, Ms. Bennett testified, "At the base of the administration policy is the fact that the Clean Air Act . . . is *already*
addressing all three of the precursor pollutants of concern [i.e.,
SO_2, NO_x, and hydrocarbons]" She conceded, "An apparent dif-
ficulty is that the Act . . . does not directly deal with secondary
products of pollution [acid sulfate and acid nitrate] that may be
experienced far from the source. Given the vigorous prevention
and control program already in place in this country, probably
the best in the world, the question before us is whether an additional program is necessary to protect against more distant effects. This is why a study program is needed, and why the Clean
Air Act gives us the confidence so that we need not move precipitously."[162]

The E.P.A. hardly moved "precipitously". In 1981, the Agency
opposed a provision that would have forced it to act on New
York's petition within four months, or face a $100,000 fine. "I am
deeply disturbed," explained Ms. Bennett, "that the general tax-
paying public would be punished for the Agency's failure to act
within the short statutory period."[6]

By 1984, nearly four *years* after New York filed its petition and
two years after the hearing was closed to further evidence, the
E.P.A. had still failed to issue a decision. In January 1984, Attorney General Abrams announced that New York would sue the
E.P.A. unless the Agency ruled on New York's petition within
sixty days. One week before the deadline was up, E.P.A. administrator William Ruckelshaus sent Mr. Abrams a letter arguing
that a lawsuit would not be productive. The state was not per-
suaded. In March 1984, New York filed suit in Federal court
charging the E.P.A. with "dereliction of a non-discretionary
duty".[163] The suit was eventually joined by all the northeastern
states, including Pennsylvania and New Jersey, as well as by the
Sierra Club, the Natural Resources Defense Council, the National Wildlife Federation, and other environmental organizations. It was the largest single lawsuit involving acid rain.

More protection to Canada

Ironically, the U.S. Clean Air Act gives more protection to Canada than it does to the United States when it comes to acid rain.[131] Section 115 of the Act requires a state to reduce emissions that "cause or contribute to" an air pollution problem "which may reasonably be anticipated to endanger public health or welfare in a foreign country." This is a remarkably broad statute, certainly much less stringent than the one that applies within the U.S. Under the Carter Administration, E.P.A. administrator Douglas Costle took the first steps in implementing Section 115. There is no question that SO_2 emissions in the U.S. "cause or contribute to" acid rain in Canada, or that acid rain "reasonably may be anticipated to endanger" Canada's welfare. However, Costle left office before determining which states are not in compliance with Section 115. Once states are notified by the E.P.A. administrator that they are out of compliance, they are required by law to reduce emissions. One obvious defect of the provision is that *every* state contributes to acid rain in Canada, since at least some SO_2 emissions from every state wind up in Canada. The provision does not provide a mechanism for determining which states are the most important polluters. This is left to the discretion of the Administrator and is clearly a political nightmare.

In the spring of 1984, New York brought suit in Federal court to force the E.P.A. to either continue the process initiated by Costle or else to issue a formal policy explaining why the provision should not be implemented.[163]

Why not sue the polluters?

At this point the reader may wonder why the northeastern states bothered at all with an E.P.A. that was apparently a reluctant guardian of air quality. If the northeastern states are confident they have a case, why don't they sue the midwestern polluters directly in Federal court? They have — but with little success. For one thing, defendants must be sued in the Federal district in which they reside — and in this case the Midwestern district courts have been generally unsympathetic to environmental issues. Furthermore, the standard of evidence in a court case is much stricter than it would be in a legislative or regulatory hearing. As one official in New York's Department of Law put it, "You have to haul into court three dead spruce trees and five dead fish and convince the judge that power plant X was directly responsible." More importantly, the midwestern district courts have been reluctant to become entangled in the technical details of acid rain, choosing instead to focus on purely procedural issues and defer to the Environmental Protection Agency on the science. The courts' deferral — some would say abdication — to the executive branch has worried those who believe that the right to sue in Federal court should not be traded for the whims of politicians in the executive branch or Congress.

These fears were borne out in 1981 when the Supreme Court ruled, 6-3, that Federal courts may not impose water pollution standards more stringent than those set by Congress in the Clean Water Act.[164] The Court's decision meant, in effect, that a state's right to sue another state under the Federal "common law of nuisance" was pre-empted by the mere existence of Congressional legislation in the area, even though the legislation might not cover the specific dispute at issue. If the same logic were applied to air pollution, it would mean that states could not sue each other over acid rain, but would have to rely on the Clean Air Act, even though the Act provides no remedy for acid rain! In a brief filed on behalf of the plaintiff, New York argued that in order to "protect the health and welfare of their citizens," states should be able to sue under the Federal common law of nuisance when Congress has not provided an adequate remedy. The Court's decision in effect puts the burden on Congress to foresee all possible contingencies when it drafts legislation governing interstate pollution. But that requires a wisdom of Congress that few think it now possesses; a good example is the failure of Congress to foresee the acid rain problem when it drafted the Clean Air Act in 1970. The Court's decision makes it all the more imperative to tighten the Clean Air Act's provisions governing interstate pollution.

A larger issue

The E.P.A. announced in August 1984 that it planned to deny New York's petition to curb pollution from midwestern sources. The Agency said that the northeastern states "have not made a persuasive technical case that the existing requirements of the Clean Air Act are being violated by interstate transport of air pollutants." New York's Attorney General, Robert Abrams, called the decision "legally distorted and scientifically dishonest" and said it "flies in the face of an avalanche of scientific evidence proving that excessive sulfur emissions from Midwestern states are responsible for billions of dollars of environmental damage in the Northeast."

Beyond the actual outcome of the case there sprawls a much larger issue. What are we to make of the spectacle of endless litigation, of years spent arguing the interpretation of ambiguous passages, of paperwork that would sink a ship? To some, the process is inspiring; it is proof that every interest is represented and every viewpoint heard. To others, it is an unmitigated waste of time, money, and energy; it is a game in which the public are the ultimate losers. The paradox remains: We resort increasingly to the courts because cooperation is a scarce commodity. Yet the opportunities for delay and deceit are so numerous and so inviting that it is doubtful real progress will be made on acid rain — or any similar issue, for that matter — unless there is some modicum of cooperation from the parties involved. Perhaps the solution is equally paradoxical. We need to strengthen and streamline our environmental laws, to close the kinds of loopholes described in this chapter. Yet at the same time we need more than ever to search for that elusive route to consensus and cooperation. It is uncharted territory.

CONCLUSION

"They hurled three shafts of twisting hail, three of raining cloud, three of ruddy lightning, and three of the south wind. Now they mixed in tremendous flashes, roarings, fear, and anger with persecuting flames."

Vulcan's workshop, in *The Aeneid.*

The ancient Romans worshipped Vulcan, the god of fire, with an unusual festival that is especially relevant to acid rain. Gathering in the northern part of the forum in late August, the heads of Roman families would toss small fish, brought by fisherman from the Tiber, into a roaring fire.[165] The sacrifice is thought to have been an appeasement to Vulcan, an offering of a creature normally impervious to fire. Today, more than two thousand years later, fish are no longer beyond Vulcan's reach. As acidic pollutants from the burning of fossil fuels find their way into lakes and streams, the earth's watery creatures are again being sacrificed on the altar of fire — but no longer is it a festival.

The evidence is now massive and convincing that acid rain and related forms of air pollution are taking a serious toll on lakes and streams, forests and soils, water supplies, air quality and human health — a toll that most Americans would find unacceptable. The wide variety of resources affected by acid rain all have one important feature in common, a feature that helps to explain why the problem has been allowed to fester for so long: The resources belong to everyone, and therefore to no one. Responsibility for these resources has been diffused so widely that it is virtually non-existent. Furthermore, the effects of acid rain have been cumulative, gradual, and for the most part, un-

123

dramatic. There are "tongues in trees," as Shakespeare told us, "and books in the running brooks," but who has heard the trees speak of acid rain? Our resources have corroded as slowly as silently as the historic statues of Paul Revere in Boston and the Embattled Farmer near the North Bridge in Concord. Even the acid haze before our very eyes is becoming accepted as a fact of life. As a nation, we seem to be galvanized only by crises, while our chronic problems slip through the institutional cracks.

The causes of acid rain are known and the remedy is clear: reduce emissions of SO_2 and NO_x. It should be emphasized that reducing these two substances will actually reduce five key air pollutants: SO_2, NO_x, acid sulfate, acid nitrate, and ozone (since NO_x contributes to the formation of ozone). The benefits of reducing these pollutants would be substantial. For example, all five pollutants are respiratory irritants, all five degrade man-made materials, and all have been implicated in damage to forests. New York State has already passed legislation to reduce its SO_2 and NO_x emissions by 30% within a decade, and Massachusetts is soon to follow suit. But so much air pollution is transported among states that no state can hope to end its acid rain alone; acid rain is a national problem, and its solution must be national in scope. Of the nearly 1,000 power plants in the United States, just 50 or so of the largest, coal-fired plants are at the core of the acid rain problem in the eastern U.S. Controlling SO_2 emissions from these 50 power plants — which emit more than one third of all airborne sulfur in the region — would significantly reduce acid rain, not just in the Northeast but throughout the eastern U.S.

Reducing SO_2 emissions will require a three-fold approach: 1) cleaning coal to remove much of the sulfur; 2) burning low-sulfur rather than high-sulfur coal, where appropriate; and 3) installing scrubbers or other pollution-control devices. Incorporating the latest pollution-control technology, including "staged combustion" techniques, would reduce not only SO_2 emissions but NO_x emissions as well. Care must be taken to limit NO_x emissions from both smokestacks and motor vehicles to prevent NO_x from offsetting the benefits of reducing SO_2.

These measures will be costly. Reduced demand for high-sulfur coal will put many miners out of work over the next decade, though new jobs will be created in the low-sulfur coal industry. The cost of installing scrubbers may be several billion dollars annually, adding about 10% to electric bills in many states. Yet at present, coal-generated electricity in the most polluting midwestern states is nearly 4 times cheaper than oil-generated electricity in the Northeast — largely because the real costs of pollution have been shifted from the coal-burners to the affected regions downwind.

It cannot be said that the public debate on acid rain has brought out the best in our leaders. The debate has seen our Environmental Protection Agency protect virtually every interest *but* the environment. It has seen a scientific community reticent to speak up when science has been distorted and scientists intimidated. A flaccid and docile press has been unwilling to ask hard questions of our leaders and to expect rational answers. The courts have thrown up their hands and deferred to a reluctant Congress only too willing to play politics. The energy industry and the Reagan Administration have stonewalled, and appear to have asked themselves only one soul-searching question, the one made famous by cartoonist William Hamilton: "Would a study shut them up?" Something else has been lost too, and we had better make sure we regain it: a sense that we are making progress, that our institutions are healthy enough and our leaders credible enough to handle the challenge of the future. For as we change the face of the earth on a global scale, as we sit on the threshold of altering the earth's very climate, the future is going to be challenge enough.

It is now more than a century since the term "acid rain" was coined. A debt is coming due, and it will surely be paid. Only the choice of currency is ours. Science has given us a stark warning, and it is time for technology and politics to pay heed, to make contact with the deepest currents that give our nation character. For we still respect ourselves enough to leave the wild parts as we found them. We still hear an ancient music when the wind sweeps across water that shelters life, that harbors mysteries. And when we look to the heavens with our aspirations, we expect to see an untainted sky.

APPENDIX

Plate 1

a) View near the summit of Camels Hump, Vermont in 1963.
   (Photo courtesy Hubert Vogelmann)

b) The same view in 1983.
   (Photo courtesy Tim Scherbatskoy)
Acid deposition is believed to have contributed to the extensive
damage to red spruce and other trees at Camels Hump.

Plate 2 View looking south from Mt. Mansfield, Vermont with the arrival of an air mass that had stagnated over the Midwest.

a) July 14, 1982. Visibility 35 miles. Sulfate concentration in the air less than 1 microgram/cu. meter.

b) July 15, 1982. Visibility 13 miles. Sulfate concentration 6 micrograms/cu. meter.

c) July 16, 1982. Visibility 6.8 miles. Sulfate concentration 13 micro-grams/cu. meter.

d) July 17, 1982. Visibility less than 5 miles. Sulfate concentration 35 micrograms/cu. meter. Rain later that night had pH 3.47.

Photos courtesy Richard Poirot, Vermont Agency of Environmental Conservation.

Chemistry of Rainfall in 19th Century Britain
(avg. concentration in microequivalents/liter)

| | Countryside | | | City | | |
	Scotland	England	Eng. Towns	Liverpool	Glasgow	U.S. (1980's)
			Britain (1860's)			
Sulfate	40	100	680	800	1400	60
Nitrate	5–8	12	14	10	40	24
Acidity	3–6	0	170	230	300	60–100
Ammonia	30–110		300	320	540	10

Source: Ref. 5.

Table 1. The Composition of Rain from Five Regions of the United States

ION	Aurora, N.Y.	Wooster, Ohio	Clemson, S. Carolina	Trout Lake, Wisconsin	Schmidt Farm, Oregon
$SO_4^=$	82.1	77.5	41.0	36.8	10.4
NO_3^-	40.1	31.0	16.6	24.5	3.8
H^+	78.	70.	39.	22.	5.
NH_4^+	23.8	20.2	9.3	22.8	2.5
Ca^{++}	10.6	11.4	4.8	11.8	3.3
Mg^{++}	3.3	3.2	2.3	2.9	3.6
pH	4.10	4.15	4.41	4.66	5.32

(Concentration in microequivalents/liter)

Source: Ref. 52

Table 2. Sulfur Emissions and Wet Sulfur Deposition in the Eastern U.S. for 1980 (As Sulfate)

State	Emissions		Wet Deposition	
	Total	Density	Total	Density
	(tonnes x 10^3)	(tonnes/sq. mi.)	(tonnes x 10^3)	(tonnes/sq. mi.)
Maine	134.2	4.0	155.6	4.68
New Hampshire	148.6	16.0	52.6	5.65
Vermont	20.5	2.1	(57.7)	(6.0)
Massachusetts	651.4	78.9	(49.5)	(6.0)
Rhode Island	43.8	36.0	(7.3)	(6.0)
Connecticut	203.3	40.6	(30.1)	(6.0)
New York	1471.2	29.7	341.4	6.88
New Jersey	493.6	63.0	(47.0)	(6.0)
Pennsylvania	2953.1	65.1	424.8	9.37
Delaware	123.8	60.2	(12.3)	(6.0)
Maryland	450.0	42.5	(63.5)	(6.0)
Virginia	445.5	10.9	189.9	4.65
West Virginia	1491.3	61.7	274.0	11.33
Kentucky	1510.9	37.4	(323.2)	(8.0)
South Carolina	430.4	13.9	190.0	6.15
North Carolina	843.4	16.0	335.0	6.37
Tennessee	1480.9	35.1	309.3	7.32
Georgia	1142.3	19.4	355.1	6.03
Mississippi	267.4	5.6	319.6	6.70
Alabama	1008.8	19.5	(309.6)	(6.0)
Florida	1273.2	21.7	206.9	3.53
Arkansas	107.6	2.0	(318.6)	(6.0)
Missouri	1839.0	38.5	(418.1)	(6.0)
Louisiana	355.3	7.3	(291.1)	(6.0)
Illinois	2090.2	37.1	423.0	7.50
Indiana	2836.1	78.1	(308.5)	(8.50)
Ohio	3821.7	92.7	425.3	10.3
Michigan	1270.5	21.8	342.8	5.89
Wisconsin	843.8	15.0	(224.6)	(4.0)
Minnesota	400.9	4.8	264.7	3.15
Iowa	469.5	8.3	(225.2)	(4.0)
TOTAL	30,628.1		7,296.3	

Source: Final Emissions Inventory by U.S./Canada Work Group 2, Report No. 2-4 (Draft), September 4, 1981; and National Atmospheric Deposition Program Network for 1980. Numbers in parentheses are estimated values, interpolated from adjoining states.

**Table 3. Top 20 Coal-Fired Power Plants in the U.S.
Ranked According to Total Sulfur Emissions in 1979 (As Sulfate)**

Rank	Plant	State	Estimated Emissions _Thousands of tonnes/year_
1	Paradise	Kentucky	558.8
2	Muskingum	Ohio	510.3
3	Gavin	Ohio	509.3
4	Cumberland	Tennessee	434.6
5	Monroe	Michigan	397.4
6	Clifty Creek	Indiana	395.6
7	Gibson	Indiana	391.7
8	Baldwin	Illinois	386.9
9	Labadie	Missouri	336.0
10	Kyger Creek	Ohio	308.3
11	Bowen	Georgia	303.9
12	Conesville	Ohio	280.2
13	Mitchell	West Virginia	279.3
14	Hatfields	Pennsylvania	251.0
15	New Madrid	Missouri	246.0
16	Sammis	Ohio	241.1
17	Wansley	Georgia	239.6
18	Homer City	Pennsylvania	238.7
19	Johnsonville	Tennessee	236.9
20	Gaston EC	Alabama	232.2
	TOTAL		6,777.2

Source: Ref. 37

Table 4. Atmospheric Sulfur Budget for the Eastern U.S.

Term	Magnitude _(Millions of tonnes sulfate/year)_
EMISSIONS:	
Man-made emissions	30.6
Biological emissions	.36
From west	1.2
From Canada to U.S.	2.1
Total:	34.3
REMOVAL:	
Wet deposition ("acid rain")	7.3
Dry deposition	9.9
From U.S. to Canada	6.0
From Canada to U.S.	11.7
Total:	34.9

Source: See text.

Table 5. The contribution to ambient sulfate in the Adirondacks from different source regions *(Mean of seven transport model calculations)*

Source Region	Contribution (Percent)
Michigan	6.57
Illinois–Indiana	10.10
Ohio	16.23
Pennsylvania	10.46
New York to Maine	9.93
Kentucky–Tennessee	3.93
W. Virginia to N. Carolina	7.34
Rest of Eastern U.S.	7.45
Ontario	19.44
Quebec	8.33
Atlantic Provinces	0.22

Source: Phase II Working Report by Work Group 2.
U.S./Canada Work Group on Transboundary Air Pollution,
Report #2–13, July 10, 1981.

Table 6. Percentage of pollutants in precipitation brought to the Adirondacks from 3 sectors

	Sulfate	Nitrate	Acidity
Ohio Valley/Midwest	64%	65%	62%
Canadian/Great Lakes	31%	28%	31%
Northeastern	5%	7%	7%

Source: Ref. 79.

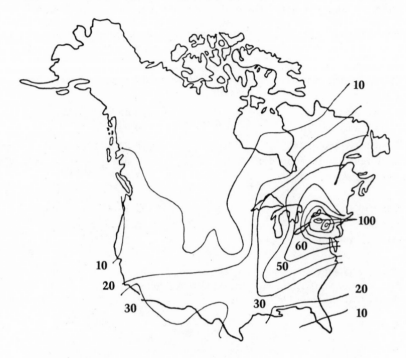

Figure 1. Average concentration of acidity in North American precipitation. (In microequivalents / liter) Data from Ref. 54.

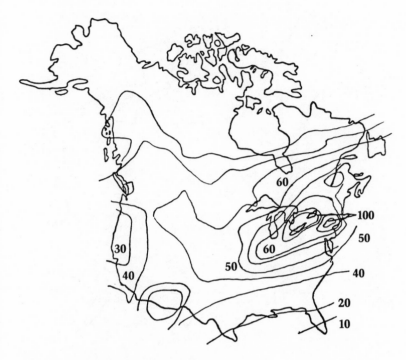

Figure 2. Average concentration of sulfate in North American precipitation. (In microequivalents / liter) Data from Ref. 54.

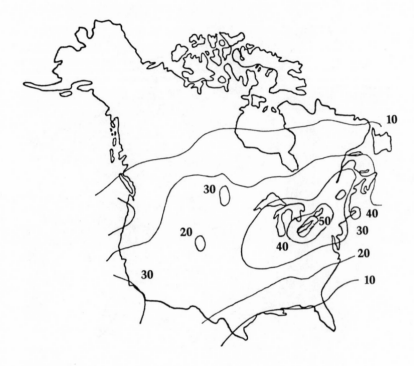

Figure 3. Average concentration of nitrate in North American precipitation. (In microequivalents/liter) Data from Ref. 54.

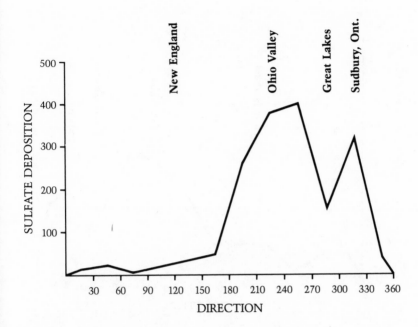

Figure 4. Sulfate deposited by rain and snow coming from twelve different sectors to Whiteface Mountain, N.Y., for 1978. (In milligrams/meter²). Data from Ref. 79.

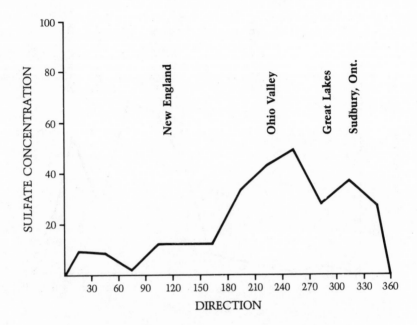

Figure 5. The average sulfate concentration in rain and snow brought to Whiteface Mountain, N.Y. by air masses from twelve sectors. (In micromoles/liter. 0-degrees is North.) Data from Ref. 79.

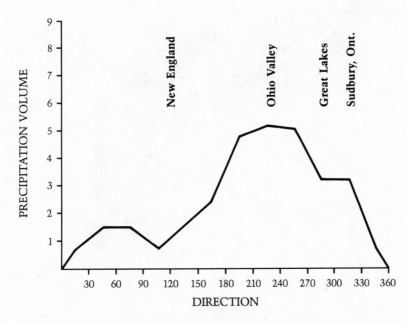

Figure 6. Precipitation brought by air masses from twelve different sectors to Whiteface Mountain, N.Y. for 1978. (Precipitation in liters.) Data from Ref. 79.

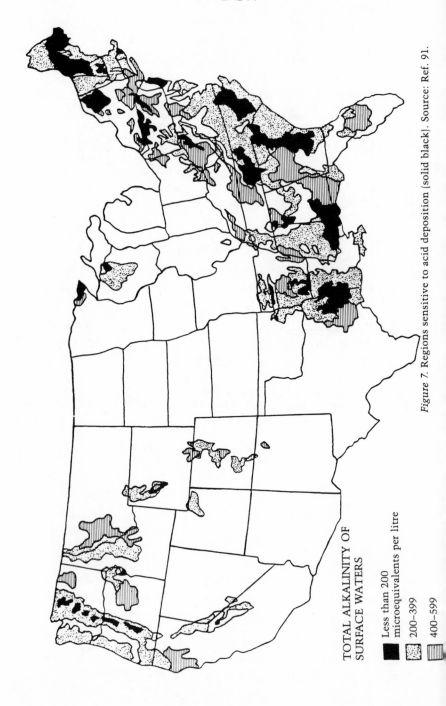

Figure 7. Regions sensitive to acid deposition (solid black). Source: Ref. 91.

TOTAL ALKALINITY OF
SURFACE WATERS

■ Less than 200
microequivalents per litre

▒ 200–399

▥ 400–599

GLOSSARY

ACID SULFATE: The chief component of acid rain and acid haze in the eastern U.S. Commonly known as sulfuric acid (H_2SO_4), it dissolves in water to form acid (H^+) and sulfate ($SO_4^=$). It is formed by the oxidation of sulfur dioxide and other forms of sulfur pollution.

ACID NITRATE: A major component of acid rain in the eastern U.S., and a principal component of acid fog in parts of California. Commonly known as nitric acid (HNO_3), it dissolves in water to form acid (H^+) and nitrate (NO_3^-). It is formed by the oxidation of nitrogen oxides.

ACID DEPOSITION: The total acid deposited to earth both in precipitation ("acid precipitation") and dry form ("dry deposition").

EASTERN U.S.: The 31 states bordering or east of the Mississippi River.

EQUIVALENT: A measure of the quantity of an acid or base. One *equivalent* of a substance is the amount that either produces or combines with one gram of hydrogen ions. For example, one equivalent of acid sulfate is 49 grams; one equivalent of carbonate is 30 grams. One equivalent of a base will exactly neutralize one equivalent of acid.

MICROEQUIVALENT: One-millionth of an equivalent.

NO_x: Refers to nitrogen oxides, which are a major cause of acid rain. The oxides are nitric oxide (NO) and nitrogen dioxide (NO_2), collectively abbreviated NO_x, where "x" stands for 1 or 2.

pH: A measure of acidity. (Literally, "power of hydrogen.") On the pH scale, 7 is neutral (pure water); a pH below 7 is acidic, greater than 7 is alkaline. A decrease of one pH unit represents a *ten-fold increase* in acidity. For example, pH 4 is one hundred times as acidic as pH 6. The pH scale is related to the unit "microequivalents per liter" as follows:

> pH 6 = 1 microequivalent acid per liter
> pH 5 = 10 microequivalents acid per liter
> pH 4 = 100 microequivalents acid per liter and so forth.

SO_2: Stands for sulfur dioxide, a major cause of acid rain. It is a colorless gas, given off during the burning of fuels containing sulfur.

TONNES: One tonne = one metric ton = one thousand kilograms. One tonne is approximately 1.1 U.S. tons.

REFERENCES

[1]Compiled by Members of the Water Board. 1868. *History of the introduction of pure water into the city of Boston from A.D. 1652 to 1850.* Boston, MA: Alfred Mudge and Son.

[2]Boucher, N. 1984. Acid stress in Quabbin area. *The Boston Globe.* May 14.

[3]Brimblecombe, P., Ogden, C. 1977. Air pollution in art and literature. *Weather.* 32:285.

[4]Brimblecombe, P. 1977. London air pollution 1500-1900. *Atmospheric Environment.* 11:1157.

[5]Smith, R.A. 1872. *Air and Rain: The Beginnings of a Chemical Climatology.* London: Longmans, Green.

[6]Bennett, K. 1981. EPA warns against 'premature action' on acid deposition. Press release. Oct. 29. Washington, DC: Environ. Prot. Agency.

[7]Overrein, L.N., Seip, H.M., Tollan, A., eds. 1980. *Acid precipitation — effects on forests and fish. Final report of the SNSF project. 1972-1980. (Sur nedbørs virkning på skog og fisk.)* Osla-Ås: Norwegian Ministry of Environ.

[8]Cavender, J.H., Kircher, D.S., Hoffman, A.J. 1973. *Nationwide air pollution emission trends, 1940-1970.* Research Triangle Park, NC: U.S. Environ. Protect. Agency.

[9]JASON Report. 1981. *The physics and chemistry of acid precipitation.* Tech. Rep. JSR-81-25. Palo Alto, Calif: SRI Int.

[10]Patrick, R., Binetti, V.P., Halterman, S.G. 1981. Acid lakes from natural and anthropogenic causes. *Science.* 211:446-48.

[11]Odén, S. 1968. *The acidification of air and precipitation and its consequences in the natural environment.* Ecology Committee Bull. No. 1. Stockholm: Swedish National Res. Council. (Arlington, VA:Translation Consultants, Ltd.).

For a detailed historical account, see Cowling, E.B. 1982. Acid precipitation in historical perspective. *Environ. Sci. Technol.* 16:110A-23A.

[12]Odén, S. 1967. *Dagens Nyheter.* Oct. 24.

[13]U.S. Environmental Protection Agency. 1982. *Critical assessment document. The acidic deposition phenomenon and its effects.* Washington, DC: U.S. Environ. Prot. Agency. 2 vols.

[14]Beamish, R.J., Harvey, H.H. 1972. *J. Fish Res. Board Can.* 29:1131-43.

[15]Schofield, C.L. 1976. *Ambio.* 5:228-30.

[16]Pfeiffer, M. H., Festa, P. J. 1980. *Acidity status of lakes in the Adirondack region of New York in relation to fish resources.* Bur. of Fisheries, Div. Fish and Wildlife. Albany, NY: NY Dept. Environ. Conserv.

[17]Boyle, R.H., Boyle, R.A. 1983. *Acid Rain.* p. 59-60. New York: Schocken.

[18]Vogelmann, H.W. 1982. Catastrophe at Camels Hump. *Natural History* November.

[19]Siccama, T.G., Bliss, M., Vogelmann, H.W. 1981. *Decline of red spruce in the Green Mountains of Vermont.* Contribution from the Vermont Agricultural Experiment Station.

[20]Scott, J.T., Siccama, T.G., Johnson, A.H. 1983. *Long-term changes in the spruce-fir forests in the Adirondacks.* Presented at the U.S. Environ. Prot. Agency Research Peer Review. Feb. 1983. Raleigh, NC. Also Raynal, D.J., Leaf, A.L., Manion, P.D., Wang, C.J.K. 1980. *Actual and potential effects of acid precipitation on a forest ecosystem in the Adirondack Mountains.* New York State Energy Res. Dev. Admin. Rep. 80-28. Albany, NY.

[21]Johnson, A.H., Siccama, T.G., Wang, D. Turner, R.S., Barringer, T.H. 1981. Recent changes in patterns of tree growth rate in the New Jersey Pinelands: a possible effect of acid rain. *J. Environ. Qual.* 10:427-30.

[22]Johnson, A.H. 1983. *Assessing the possibility of a link between acid deposition and recent changes in forest growth.* Presented at U.S. Environ. Prot. Agency Research Peer Review. Feb. 1983. Raleigh, NC.

See also: Johnson, A.H., Siccama, T.G. 1983. Acid deposition and forest decline. *Environ. Sci. Technol.* 17:294A-305A.

[23]Taylor, F., Taylor, J. A., Symons, G. E., Collins, J. J., Schock, M. 1984. Acid precipitation and drinking water quality in the eastern United States. Drinking Water Res. Div., Mun. Environ. Res. Lab., Coop. Agreem. No. CR807808010. Cincinnati, OH: U.S. Environ. Prot. Agency.

[24]Letters from G.W. Fuhs (Dir. Environ. Health Cent., NY State Dept. Health) to the Hon. T. Moffett (Chairman, Environ., Energy and Nat. Resources Subcomm., U.S. House Rep.) July 16, 1980, and to M.H. Surgan (Environ. Protection Bureau, NY State Dept. of Law) July 15, 1981. Reproduced in Abrams, R. 1981. *Attachments to evidence summary.* Before the U.S. Environmental Protection Agency Section 126 Hearing on Interstate Transport of Pollution. Docket No. A-81-09. Albany, NY.

[25]Bloomfield, J.A., Quinn, S.O., Scrudata, R.J., Long, D., Richards, A., et al. 1980. Atmospheric and watershed impact of mercury to Cranberry Lake, St. Lawrence County, New York. In *Polluted Rain,* ed. Toribara, Miller, Morrow. New York: Plenum.

Clarkson, T.W. 1975. *Exposure to methylmercury in Grassy Narrows and White Dog Reserves.* Interim Report. Medical Services Branch. Ottawa: Ministry Health and Welfare.

[26]U.S./Canada Memorandum of Intent on Transboundary Air Pollution. Feb. 1983. Executive Summaries Work Group Reports. Washington, D.C.

[27]Office of Technology Assessment. 1982. *The Regional Implications of Transported Air Pollutants: An Assessment of Acidic Deposition and Ozone* Interim Draft. Washington, DC: OTA.

[28]Reynolds, L. 1982. Economic and employment impacts of proposed acid rain legislation. Testimony Before the U.S. Senate Foreign Relations Comm., Subcomm. on Arms Control, Oceans, International Operations and Environment. Feb. 10, 1982. Washington, DC.

[29]Raasch, C. 1984. Coal rates: old battle heats up. *USA Today.* July 12. p. B1.

[30]Mares, J. 1982. Testimony Before the U.S. Senate Comm. on Energy and Natural Resources. Aug. 19, Pub. No.97-87. Washington, DC. p. 250.

[31]Kearney, J. 1982. *Ibid.* p. 1047.

[32]Caulfield, C. 1984. Britain's bill for stopping acid rain is halved. *New Scientist.* May 17. p. 6.

[33]Likens, G., Butler, T.J. 1981. Recent acidification of precipitation in North America. *Atmos. Environ.* 15:1103-9.

[34]Lewis, W.M., Jr., Grant, M.C. 1980. Acid precipitation in the Western U.S. *Science* 207:176-7.

[35]Galloway, J.H., Likens, G.E. 1981. Acid precipitation: the importance of nitric acid. *Atmos. Environ.* 15:1081-5.

[36]Rom, W.N., Lee, J. 1983. Energy alternatives: what are their possible health effects? *Environ. Sci. Technol.* 17:132A-44A.

[37]Canadian Govt. Standing Comm. Fisheries and Forestry. 1981. *Still Waters.* (Rep. Subcomm. on Acid Rain) Ottawa.

[38]National Academy of Sciences. 1981. *Atmosphere-biosphere interactions: Toward a better understanding of the ecological consequences of fossil fuel combustion.* Washington, DC: NAS/Natl. Res. Counc.

[39]Ref. 13, Vol. 2, pp. 4-94.

[40]Reinhold, R. 1982. Acid rain issue creates stress between administration and science academy. *The New York Times.* June 8.

[41]Office of Science and Technology Policy. 1983. Summary Remarks on Acid Rain. Acid Rain Peer Review Panel. Executive Office of the President. June 27, 1983. Washington, DC.

[42]Kranish, M. 1984. Acid rain report said suppressed. *The Boston Globe.* Aug. 18. p. 1.

[43]Shabecoff, P. 1983. Acid rain options to be listed soon. *The New York Times.* Sept. 1, 1983.

[44]News Article. 1983. E.P.A. study group may urge 50% cut in acid rain emissions. *The New York Times.* July 30.

[45]Oberdorfer, D. 1983. Ruckelshaus tells Canada action put off on acid rain. *The Boston Globe.* Oct. 17, 1983. Shabecoff, P. 1983. Ruckelshaus puts off plan to curb acid rain. *The New York Times.* Oct. 23, 1983.

[46]Shabecoff, P. 1984. Toward a clean and budgeted environment. *The New York Times.* Oct. 2. p. A28.

[47]Trausch, S. 1983. E.P.A. accused of weakening acid rain rule. *The Boston Globe.* Nov. 15. p. 3.

[48]Sangeorge. R. 1983. U.S. court orders new acid rain regulations. *The Boston Globe.* Oct. 12.

[49]Readers' poll. 1984. *Chemical Processing.* July. p. 22-3.

[50]Ward, B. 1984. The intractable politics of acid rain. *Pollution Engineering.* June. pp. 20-21.

[51]See Ref. 17, p. 96.

[52]Natl. Atmos. Deposition Prog. 1981-2. *Data Report: Precipitation Chemistry, 1980.* Natural Resource Ecol. Lab. Fort Collins, Colo: Colorado St. Univ. 4 vols.

[53]MAPS3S/RAINE Research Community. 1982. The MAP3S/RAINE precipitation chemistry network: statistical overview for the period 1976-80. *Atmos. Environ.* 16:1603-32.

[54]Munger, J.W. Eisenreich, S.J. 1983. Continental-scale variation in precipitation chemistry. *Environ. Sci. Technol.* 17:32A-42A.

[55]Ferek, R. 1982. *A study of aerosol acidity over the Northeastern United States.* Cooperative Thesis No. 68. Boulder, Colo: Natl. Cent. for Atmos. Res. and Florida St. Univ.

[56]Waldman, J.M., Munger, J.W., Jacob, D.J., Flagan, R.C., Morgan, J.J., et al. 1982. Chemical composition of acid fog. *Science.* 218:677-80.

[57]Szabo, M. F., Spaite, P. W. 1982. *Acid rain: commentary on controversial issues and observations on the role of fuel burning.* (Prep. for U.S. Dept. Energy) Rep. DOE/MC/19170-1168. Morgantown, WVa: Morgantown Energy Tech. Cent.

[58]Harte, J. 1983. An investigation of acid precipitation in Qinghai province, China. *Atmos. Environ.* 17:403-8.

[59]Delmas, R., Boutron, C. 1978. Sulfate in Antarctic snow: spatio-temporal distribution. *Atmos. Environ.* 12:723-8.

[60]Herron, M. M., Langway, C. C., Weiss, M. V., Cragin, J. H. 1977. Atmospheric trace metals and sulfate in the Greenland ice sheet. *Geochim. Cosmochim. Acta.* 41:915-20.

[61]Logan, J. A. 1983. Nitrogen oxides in the troposphere: global and regional budgets. *J. Geophys. Res.* 88:10785-807.

[62]Calvert, J.G., Stockwell, W. R. 1983. Acid generation in the troposphere by gas-phase chemistry. *Environ. Sci. Technol.* 17:428A-43A.

[63]National Academy of Sciences. 1983. *Acid deposition: Atmospheric processes in eastern North America.* Washington, DC: NAS/Natl. Res. Counc.

[64]U.S./Canada Memorandum of Intent on Transboundary Air Pollution. 1981. *Final Emissions Inventory.* Rep. 2-4. (Work Group 2) Washington, DC: U.S. Environ. Prot. Agency.

[65]Galloway, J. N., Whelpdale, D. M. 1980. An atmospheric sulfur budget for North America. *Atmos. Environ.* 14:409-17.

[66]Rice, H., Nochumson, D. H., Hidy, G. M. 1981. Contribution of anthropogenic and natural sources to atmospheric sulfur in parts of the United States. *Atmos. Environ.* 15:1-9.

[67]Adams, D. F., Farwell, S. O., Robinson, E., Pack, M. R., Bamesberger, W. L. 1981. Biogenic sulfur source strengths. *Environ. Sci. Technol.* 15:1493-8.

[68]Hofmann, D. J., Rosen, J. M. 1983. Sulfuric acid droplet formation and growth in the stratosphere after the 1982 eruption of El Chichon. *Science.* 222:325-7.

[69]Arnold, F., Buhrke, T. 1983. New H_2SO_4 and HSO_3 measurements in the stratosphere — evidence for a volcanic influence. *Nature.* 301:293-6.

[70]Dana, M. T. 1980. Sulfur dioxide versus sulfate wet deposition in the eastern U.S. *J. Geophys. Res.* 85:4475-80.

[71]Judeikis, H. S., Stewart, T. B., 1976. Laboratory measurements of sulfur dioxide deposition velocities onto selected building materials and soils. *Atmos. Environ.* 10:769-76.

[72]McMahon, T. A., Denison, P. J. 1979. Empirical atmospheric deposition parameters: a survey. *Atmos. Environ.* 13:571.

[73]Jickells T., Knap, A., Church, T., Galloway, J., Miller, J. 1982. Acid rain on Bermuda. *Nature.* 297:55-7.

[74]Wolff, G. T., Kelly, N.A., Ferman, M. A. 1981. On the sources of summertime haze in the eastern U.S. *Science.* 211:703-5.

[75]Gillani, N. V., Shannon, J. D., Patterson, D. E. 1983. Transport processes. In Ref. 13, Vol. 1.

[76]U.S./Canada Memorandum of Intent on Transboundary Air Pollution. 1982. *Atmospheric Sciences and Analysis.* Rep. 2F (Work Group 2) Washington, DC.

[77]Wolff, G. T. 1982. Acid precipitation. *Science.* 216:1172.
See also Samson, P. J. 1978. Ensemble trajectory analysis of summertime sulfate concentrations in New York State. *Atmos. Environ.* 12:1889-93.

[78]U.S./Canada Memorandum of Intent on Transboundary Air Pollution. 1981.

Phase II Working Report. Rep. 2-13. (Work Group 2) Washington, DC.

[79]Wilson, J. W., Mohnen, V. A., Kadlecek, J. A. 1980. *Wet deposition in the northeastern United States.* Atmos. Sci. Res. Center Pub. 796. Albany: St. Univ. of NY.
  See also data in Ref. 63.

[80]Vidulich, G. A. and D. A. MacKoul. 1984. Chemistry of acid precipitation in central Massachusetts from 1976 to 1983. Presented at the Northeast States Acid Precipitation Symposium, March 27 and 28. Boston, MA.

[81]Poon, C. P. C. 1984. Acid precipitation: source and effects on water quality. *Ibid..*

[82]Kurtz, J. Scheider, W. A. 1981. An analysis of acidic precipitation in south-central Ontario using air parcel trajectories. *Atmos. Environ.* 15:1111-6.

[83]Thurston, G. D., Spengler, J. D., Samson, P. J. 1982. *An assessment of relationships between regional pollution transport and trace elements using wind trajectory analysis.* Presented at Specialty Conf. on Receptor Models Applied to Contemporary Pollution Problems. Oct. 17 Danvers, MA.

[84]Parekh, P. P., Husain, L. 1981. Trace element concentrations in summer aerosols at rural sites in New York and their possible sources. *Atmos. Environ.* 15:1717-25.

[85]Rahn, K., Lowenthal, D. H., Lewis, N. F. 1982. *Elemental tracers and source areas of pollution aerosol in Narragansett, Rhode Island.* Tech. Rep. Grad. School Oceanography. Kingston, RI: Univ. of Rhode Island. 119 pp.

[86]Thurston, G. D., 1983. *A source apportionment of particulate air pollution in metropolitan Boston.* Doctoral Thesis. Cambridge, MA: Harvard Univ.

[87]Koerner, R. M., Fisher, D. 1982. Acid snow in the Canadian high Arctic. *Nature.* 295:137-40.

[88]Rahn, K. A. 1981. The Mn/V ratio as a tracer of large-scale sources of pollution aerosol for the Arctic. *Atmos. Environ.* 15:1457-64.

[89]Miller, J., Yoshinaga, A. M. 1981. The pH of Hawaiian precipitation: a preliminary report. *Geophys. Res. Let.* 8:779-82.
  See also Parrington, J. R., Zoller, W. H., Aras, N. K. 1983. Asian dust: seasonal transport to the Hawaiian Islands. *Science.* 220:195-7.

[90]Hofmann, D. J., Rosen, J. M. 1980. Stratospheric sulfuric acid layer: evidence for an anthropogenic component. *Science.* 208:1368-70.

[91]Galloway, J. N., Anderson, D. S., Church, M. R., et al. 1983. Effects on aquatic chemistry. In Ref. 13, vol. 2.

[92]Lefohn, A. S. 1982. Comments on the status of knowledge concerning the effects of acidic precipitation on the aquatic and terrestrial ecosystem. Hearing Before the U.S. Senate Comm. on Energy and Natural Resources. Aug. 19, Pub. No. 97-87. Washington, DC. pp 211-47.
  Also Semonin, R. G. 1982. North American acidic precipitation issues. *Ibid.* pp. 951-67.
  Also Kramer, J., Tessier, A. 1982. Acidification of aquatic systems: a critique of chemical approaches. *Environ. Sci. Technol.* 16:606-15.

[93]McColl, J. G. 1981. Increasing hydrogen-ion activity of water in two reservoirs supplying the San Francisco Bay area, California. *Water Resources Research.* 17:1510-16.
  But cf. Bradford, G. R., Page, A. L., Straughan, I. R. 1981. Are Sierra lakes becoming acid? *California Agriculture.* 35:6-7.

[94]Krug, E. C., Fink, C. R. 1983. Acid rain on acid soil: a new perspective. *Science.* 221:520-25. For a rebuttal, see Letters. 1984. *Science.* 225:1424-34.

[95]Havas, M., Hutchinson, T. C., Likens, G. E. 1984. Red herrings in acid rain research. *Environ. Sci. Technol.* 18:176A-85A.

[96]Haines, T. A. 1981. Acid precipitation and its consequences for aquatic ecosystems. A review. *Trans. Amer. Fish. Soc.* 110:669-707. See also Ref. 13.

[97]Fromm, P. O. 1980. A review of some physiological and toxicological responses of freshwater fish to acid stress. *Environ. Biol. Fish.* 5:79-93.

[98]Baker, J. P., Schofield, C. L. 1980. Aluminum toxicity to fish as related to acid precipitation and Adirondacks surface water quality. *Proc. Int. Conf. Ecological Impacts of Acid Precipitation*, ed. Drablos, A. Tollan. SNSF Project. Oslo: Norwegian Ministry of Environ.

[99]Norton, S. A., Hanson, D. W., Campana, R. J. 1980. The impact of acid precipitation and heavy metals on soils in relation to forest ecosystems. *Proc. Symp. Effects of Air Pollution on Mediterranean and Temperate Forest Ecosystems.* June, 1980. Rep. PSW-43. pp 152-7. Riverside, Calif: U.S. Forest Service.

Also Johnson, D. W., Turner, J., Kelly, J. M. 1982. The effects of acid deposition on forest nutrient status. *Water Resources Research.* 18:449-61.

Also Budd, W. W., Johnson, A. H., Huss, J. B., Turner, R. S. 1981. Aluminum in precipitation, streams and shallow groundwater in the New Jersey Pine Barrens. *Water Resources Research.* 17:1179-83.

Also Cowling, E. B., Davey, C. B. 1981. Acid precipitation. Basic principles and ecological consequences. *Pulp and Paper.* 55:182-5.

See also Refs. 13 and 38.

[100]Whitney, H. 1984. Structural and compositional changes in a high elevation spruce forest. Presented at the Northeast States Acid Precipitation Symposium. March 27 and 28. Boston, MA.

[101]Johnson, A. H. 1979. Evidence of acidification of headwater streams in the New Jersey Pinelands. *Science.* 206:834-5.

[102]Baes III, C. F., McLaughlin, S. B. 1984. Trace elements in tree rings: evidence of recent and historical air pollution. *Science.* 224:494-7.

[103]Shabecoff, P. 1984. Damage to trees reported severe. *The New York Times.* Apr. 15, 1984.

[104]Postel, S. 1984. Air pollution, acid rain, and the future of forests. Worldwatch Paper 58. March. Washington, DC: Worldwatch Inst.

[105]Dunnett, J. S. 1983. Ozone named as culprit. *Nature.* 301:275.

[106]Cowling, E.B. 1984. What is happening to Germany's forests? *The Environmental Forum.* May. pp. 6-11.

[107]Ulrich, B., Mayer, R., Khanna, R.K. 1980. Chemical changes due to acid precipitation in a loess-derived soil in central Europe. *Soil Sci.* 130:193-9.

[108]Evans, L. S. 1982. Biological effects of acidity in precipitation on vegetation: a review. *Environ. Exper. Bot.* 22:155-69.

[109]Chang, F. H., Alexander, M. 1983. Effect of simulated acid precipitation on algal fixation of nitrogen and carbon dioxide in forest soils. *Environ. Sci. Technol.* 17:11-3.

[110]McFee, W. W., Adams, F., Cronan, C. S., et al. Effects on soils systems. In Ref. 13. vol. 2.

[111]In some soils, nitrate is not very mobile because it is quickly taken up as a nutrient by plants. Although this uptake temporarily removes acid nitrate from the soil, eventually the plants die and return acid nitrate to the soil. Plants will only be a *net* sink for acidity if there is an increase in vegetation from one year to the next — a "greening" — a trend that could not continue indefinitely. Some soils can also bind sulfate ions, but eventually these soils will become saturated. Although new soil is continuously generated by the gradual weathering of rocks and minerals and the remains of organisms, the birth of new soil is a very slow process — probably too slow to compete with the steady input of acid rain.

[112]Davis, H. 1983. Acid rain, aluminum link found. *The Washington Post.* May 24, 1983. p. A9.

Also Maugh II, T. H. 1984. Acid rain's effects on people assessed. *Science.* 226:1408-1410.

[113]Heit, M., Tan, Y., Klusek, C., Burke, J. C. 1981. Anthropogenic trace elements and polycyclic aromatic hydrocarbon levels in sediment cores from two lakes in the Adirondacks acid lake region. *Water Air Soil Poll.* 15:441-64.

[114]Clarkson, T. W., Baker, J. P., Sharpe, W. E. 1983. Indirect effects on health. In Ref. 13, vol. 2.

[115]Wheatley, B. 1979. *Methyl mercury in Canada.* Medical Services Branch, Ottawa: Ministry Health and Welfare.

[116]Jernelov, A. 1980. The effects of acidity on the uptake of mercury in fish. In *Polluted Rain,* ed. Toribara, Miller, Morrow. New York: Plenum.

[117]Ware, J. H., Thibodeau, L. A., Speizer, F. E., Colome, S., Ferris, B. G., Jr. 1981. Assessment of the health effects of atmospheric sulfur oxides and particulate matter: evidence from observational studies. *Environ. Health Persp.* 41:255-76.

[118]Wilson, R. Colome, S. D., Spengler, J.D., Wilson, D. G. 1980. *Health Effects of Fossil Fuel Burning.* Cambridge, MA: Ballinger.

[119]See Ref. 17. p. 83.

[120]Organ. for Econ. Coop. and Dev. 1981. *The Costs and Benefits of Sulfur Oxide Control.* Paris: OECD.

[121]Stevens, R. K., Dzubay, T. B., Shaw, R. W. Jr., McClenny, W. A., Lewis, C. W., et al. 1980. Characterization of the aerosol in the Great Smoky Mountains. *Environ. Sci. Technol.* 14:1491.

[122]Poirot, R., Wishinski, P. 1984. Factors contributing to visibility degradation in northern Vermont. Presented at the Northeast States Acid Precipitation Symposium. Mar. 27 and 28. Boston, MA.

[123]Keller, R. 1982. Mass. Div. Fish and Wildlife, personal communication.

[124]Blake, L. M. 1981. Liming acid ponds in New York. *N.Y. Fish and Game Journal.* 28:208-14.

Also De Pinto, J. V., Edzwald, J. K. 1983. *An evaluation of the recovery of Adirondacks acid lakes by chemical manipulation.* Rep. PB83-108498. Washington, DC: Natl. Tech. Inform. Serv.

[125]Swedish Gov., Ministry of Agriculture. *Acidification Today and Tomorrow.* Prep. for the 1982 Stockholm Conf. on Acidification of the Environ. Environment '82 Comm. Stockholm.

Also Rodhe, A. 1981. Reviving acidified lakes. *Ambio.* 10:195-6.

[126]Martin, L. R., Damschen. D. E. 1981. Aqueous oxidation of sulfur dioxide by hydrogen peroxide at low pH. *Atmos. Environ.* 15:1615-21. See also Ref. 62.

[127]Likens, G. E., Bormann, F. H., Pierce, R. S., Eaton, J. S., Johnson, N. M. 1977. *Biogeochemistry of a Forested Ecosystem.* New York: Springer-Verlag.

[128]Eldred, R. A., Ashbaugh, L. L., Cahill, T. A., Flocchini, R. G., Pitchford, M. L. 1981. The effect of the 1980 copper smelter strike on air quality in the Southwest. In Ref. 24, Attachment E.

[129]Liu, Y.A., ed. 1982. *Physical Cleaning of Coal.* New York: Marcel Dekker.

[130]Marcus, S. J. 1983. Acid rain and pollution curbs. *The New York Times.* Nov. 7, 1983. p. D1.

[131]Wetstone, G. S., Rosencranz, A. 1983. *Acid Rain in Europe and North America.* Washington, DC: Environ. Law Inst.

[132]Wilson, C. L. 1980. *Coal — Bridge to the Future. Report of the World Coal Study.* Cambridge, MA: Ballinger. Also Fennelly, P. F. 1984. Fluidized bed combustion. *Amer. Sci.* 72:254-61.

[133]Acid precipitation: a utility industry view. 1981. From an informational packet distributed by the Edison Electric Institute. Washington, DC: Edison Elec. Inst.

[134]Curtis, C. "What we know about acid rain." *Ibid.* p. 33.

[135]Bennett, K. 1982. Testimony before the U.S. Senate Committee on Energy and Natural Resources. In *Acid precipitation and the use of fossil fuels. Hearing Before the Committee on Energy and Natural Resources.* U.S. Senate. Aug. 19, 1982. Pub. No. 97-87. p. 535. Washington, DC: U.S. Senate.

[136]Funkhouser, R. 1983. Ask about acid rain. *The New York Times.* Aug. 18.

[137]News item. 1981. *Environ. Sci. Technol.* 15:11.

[138]See Ref. 17, p. 116-8 for discussion.

[139]Wingate, P. J. 1982. Wondering and worrying about acid rain. *The Wall Street Journal.* May 18.

[140]Perhac, R. Acid rain's identification crisis. p. 45. Reprinted in informational packet distributed by the Edison Electric Institute. Washington, DC: Edison Elec. Inst.

[141]Cohen, P. 1982. In Ref. 135. p. 565.

[142]Edison Electric Institute. 1982. In Ref. 135. pp. 60-75.

[143]Mohnen, V. 1981. Testimony Before the U.S. Environ. Prot. Agency Sec. 126 Hearing on Interstate Transport of Pollution. Docket No. A-81-09. June 19, 1981.

[144]Mohnen, V. 1981. Testimony Before the U.S. House of Rep. Subcomm. on Health and Environ. of the Comm. on Energy and Commerce. June 22, 1981.

[145]Lyons, R. D. 1982. 4,867 feet high in the Adirondacks, scientists study acid rain. *The New York Times.* July 30.

[146]Bird, D. 1983. Acidity in lakes attributed to natural chemicals. *The New York Times.* Mar. 17. p. B20.

[147]Tape cassette. 1981. Acid rain. NPR Journal Series. NJ-820222.01/01-C. Washington, DC: Natl. Pub. Radio.

[148]Mohnen, V. 1982. Testimony Before the U.S. Senate Committee on Environment and Public Works. May 25, 1982.

[149]Mohnen, V. 1982. Submissions to the hearing record. Before the U.S. Environ. Prot. Agency. Sec. 126 Hearing on Interstate Transport of Pollution. Docket No. A-81-09. Feb. 12, July 2, and thereafter.

[150]Confronting the acid test. 1983. *Time*. July 11. p. 19.

[151]Mohnen, V. 1983. Letter to the Editor. *Time*. Aug. 15, p. 4.

Also Mohnen, V. 1983. Letter to the Editor. *The New York Times*. July 29.

[152]Lugar, R. 1983. To combat acid rain. *The New York Times*. Aug. 15.

[153]Rahn, K. 1981. Elemental tracers and sources of atmospheric acidity for the Northeast. A statement of new evidence. Nov. 23. Letter distributed to members of Congress by K. Rahn, School of Oceanography, University of Rhode Island. Narragansett, RI.

[154]Kerr, R. A. 1982. Tracing sources of acid rain causes big stir. *Science*. 215:881.

[155]Rahn, K. 1982. Testimony Before the U.S. Senate Committee on Environment and Public Works. May 25.

[156]U.S./Canada Memorandum of Intent on Transbounding Air Pollution. 1982. Final Report. Work group 2. November. pp. 8-9.

[157]Abelson, P. 1983. Acid rain. *Science*. 221:115.

[158]Abrams, R. 1980. Petition of the State of New York for disapproval of proposed revision of state implementation plan and comments. Before the U.S. Environmental Protection Agency Section 126 Hearing on Interstate Transport of Pollution. Docket No. 5A-80-3.

[159]Proposed rules. 1981. *The Federal Register*. 46: 24602. May 1.

[160]Strelow, R. Statement on behalf of Ohio Power Company et al. 1981. Before the U.S. Envir. Prot. Agency Section 126 Hearing on Interstate Transport of Pollution. Docket No. A-81-09. June 18.

[161]*Proceedings of the New York State Acid Rain Conference*. Held Mar. 9 and 10, 1982. New York, NY: NY Dept. Environ. Conserv.

[162]Bennett, K. 1982. See Ref. 135. p. 534.

[163]Wooley, D. 1984. Multi-state legal effort to force EPA to address acid rain controls. Presented at the Northeast States Acid Precipitation Symposium, Mar. 27 and 28. Boston, MA.

[164]Greenhouse, L. 1981. Justices, 6-3, limit role of courts in imposing water pollution rules. *The New York Times*. Apr. 29.

[165]Vulcan. 1963. *Encyclopedia Britannica*. 23:262.

Roy Gould recently completed a three year scientific review and policy analysis of acid rain while a research fellow at the Harvard School of Public Health (1980–1983). He received his Ph.D. in biophysics from Harvard University in 1974. Dr. Gould has testified on acid rain before the U.S. Senate Environment Committee and the Senate Energy Committee, and he has lectured widely on acid rain before groups as diverse as the Air Force Geophysics Laboratories and the Appalachian Mountain Club. Dr. Gould maintains an active interest in the public understanding of science. In 1977–78, he was associate producer at the NOVA Science Series (WGBH-TV, Boston), and he is currently preparing an exhibit on water resources for Boston's Museum of Science.